THE
WILDLIFE
POND
BOOK

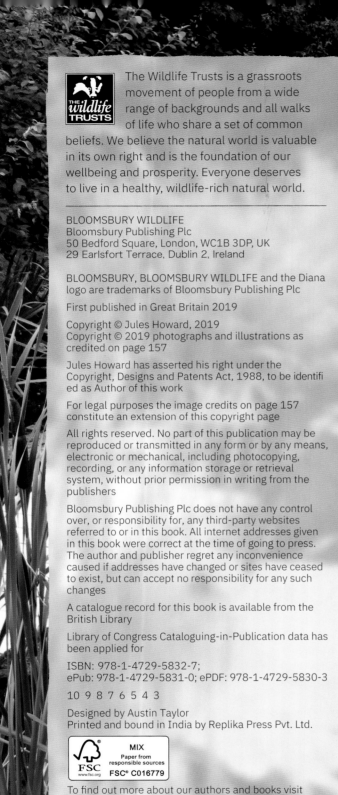

BLOOMSBURY WILDLIFE
Bloomsbury Publishing Plc
50 Bedford Square, London, WC1B 3DP, UK
29 Earlsfort Terrace, Dublin 2, Ireland

BLOOMSBURY, BLOOMSBURY WILDLIFE and the Diana logo are trademarks of Bloomsbury Publishing Plc

First published in Great Britain 2019

A catalogue record for this book is available from the British Library

Library of Congress Cataloguing-in-Publication data has been applied for

ISBN: 978-1-4729-5832-7;
ePub: 978-1-4729-5831-0; ePDF: 978-1-4729-5830-3

10 9 8 7 6 5 4 3

Designed by Austin Taylor
Printed and bound in India by Replika Press Pvt. Ltd.

FSC
www.fsc.org

MIX
Paper from responsible sources
FSC® C016779

To find out more about our authors and books visit www.bloomsbury.com and sign up for our newsletters

THE
WILDLIFE
POND
BOOK

Create Your Own Pond
Paradise for Wildlife

JULES HOWARD

BLOOMSBURY WILDLIFE
LONDON · OXFORD · NEW YORK · NEW DELHI · SYDNEY

CONTENTS

1

THE IDEAS STAGE

2

PREPARING YOUR POND

3

MANAGING YOUR POND

FOREWORD BY KATE BRADBURY

When Jules sent me his book I read it cover to cover – I was enthralled and hung on every word. It gives us a fascinating insight into the origins of ponds and their role in history – both in the British Isles but also further afield. I'd never considered 'footprint ponds' before I read Jules's book. I'd never considered prehistoric frogs and toads relying on the feet of dinosaurs to enable them to breed. But now I know. And, well, there's no going back.

A pond is surely the most joyous of garden wildlife habitats. I've lost hours staring at them – at house sparrows taking it in turns to bathe in the shallows, at common darter dragonflies dropping eggs into the pondweed, at the great fat herring gull that has taken to my new pond with such gusto, it has at least four daily swims. Ponds bring light and life to a garden. They're noisy with partying amphibians in spring and busy with flitting insects and hungry bats on summer evenings. Sit quietly next to your pond at dusk and you may even spot a hedgehog drinking at the water's edge.

It's just as exciting beneath the surface. Find your inner child and treat yourself to a fishing net and a white tray. Empty the net into the tray, and you'll come face to face with greater diving beetles, dragonfly and damselfly nymphs, freshwater shrimps and stickleback fish. And then there are all the other weird, other-worldly things that barely look alive. Have you heard of water hog-lice, copepods or pea mussels? You're about to.

Ponds are important, too. I can't think of anything that is as fun, yet as vital, as a garden pond. While we lose rivers, streams and ponds in the wild, our gardens can help mitigate some of these wildlife losses, although, tragically, not all of them. We can't provide habitats for water voles or kingfishers in our gardens. We're unlikely to provide homes for rare beetles and natterjack toads. But we can provide somewhere for common frogs and toads and palmate newts to breed, for several dragonfly species to lay eggs and for countless birds and small mammals to drink. Garden ponds provide a habitat for wildlife that would otherwise not have it, it's as simple as that. And doing so is wonderful. We should all dig a pond, and Jules is the perfect person to show us how.

I don't know how my new garden pond compares with that of a dinosaur footprint, but I'd like to think it's a suitable alternative. I dug it just two weeks ago, during a cold spring. But so far I've seen house sparrows, blackbirds, starlings, woodpigeons and collared doves bathing and drinking from it. At night I've captured hedgehogs drinking from it on my camera trap. Then there's my big fat swimming herring gull, of course. I'm driving myself mad looking for the first invertebrates to turn up; anxious for greater diving beetles, non-biting midges and pond skaters. And I'm spending far too much time wondering what's going to be first to land on my bespoke dragonfly perch.

Jules's book is a fantastic resource for wildlife gardeners and a lifeline for all who rely on ponds for breeding and survival. Why? Because you're going to read this and want to dig a pond. Even if you already have one, you're going to want to dig another. Or you might lend Jules's book to a neighbour or friend, and encourage them to dig a pond. Simply by writing this book, Jules has improved the fate of thousands of wildlife species, because his enthusiasm is so infectious he'll have us all digging holes in the garden before we've finished reading. So, what are you waiting for? Make a cup of tea, have a read, and then grab your spade. The water hog-lice are waiting...

Kate Bradbury

Award-winning author and journalist, specialising in wildlife gardening

INTRODUCTION

THE WONDER OF PONDS

LOOKING AT EARTH FROM SPACE, aliens from another planet would quickly notice the human fascination with water. They would spot the elegant ornamental ponds in front of our country houses. They would see us congregating around urban ponds in local parks on warm days. They would observe kids jumping in puddles. Pooh sticks in local streams. The familiar kidney-shaped micro-lakes that dot the nation's gardens. It would be obvious to them, I hope, that humans are a warm and mostly friendly species – a species coming to understand just how important ponds are for enriching landscapes with wildlife.

These simple habitats really do bolster local nature, as you will discover in this book. In fact, the single biggest thing you can do for local nature is to find a suitable location and reach for the spade. If we put wildlife ponds in industrial areas, in public parks, in our schools and gardens, if we put them in our shopping centres or even on our roundabouts we would make a gear-changing impact on the survival prospects of birds, beetles, bats, frogs and toads, hedgehogs (*Erinaceus europaeus*) and dragonflies. Their lives would be better for it. And our lives would be better for it too.

But ponds have other uses. Increasingly, scientists are coming to understand that ponds are a potential weapon against climate change, playing an almost unparalleled role in capturing carbon and storing it away. Ponds can be purifiers of pollution too. And ponds are a unique and accessible habitat that can provide important learning opportunities for young (and older) wannabe scientists or amateur naturalists.

In this book, you will find out more about why ponds are important and, crucially, I hope you'll find a pond that works for you. Because, genuinely, no matter how small your outside space, there are opportunities for a life-giving pond or pool somewhere within it. It is a myth that ponds have to be big to be successful; thirsty birds, colonising crustaceans and inquisitive amphibians will even use a pond the size of a washing-up tub at certain times of the year. It is also a myth that ponds have to be well tended or use pumps and filters. Another myth is that they need planting up with the finest and most expensive plants. Or that all ponds should last for decades or even centuries. Or that algae require killing with chemicals. In my years of designing

What exactly is a pond?

Freshwater scientists use the word 'pond' to describe water bodies that are between $1m^2$ in size at their smallest and 2ha in size at their largest. This definition also requires that a water body of these dimensions holds water for at least four months of the year. In this book, we have pushed this formal definition a little by occasionally including reference to smaller pools and even large puddles, which can be valuable habitats in their own right. Small pools like these can do an important service for local wildlife, including in very urban areas.

and digging ponds I have not once had to rely on such ideas, and nor should you.

If this book has one aim, therefore, it is to challenge common myths about ponds and to convey to readers that ponds really are a beautiful, firework-like habitat – a lit flame that

leads to energetic explosions of life within neighbourhoods. A habitat that, if well designed, needs little management and that is often transitory in its nature. A life-giving oasis to thousands of species. A place for fun. A place for learning. A place for memories and first moments. Ponds are, quite simply, not akin to any other habitat I can think of. In fact, I like to think that aliens from space could quickly come to understand the qualities of a world-changing species like our own by assessing how many ponds we have per square mile. This is humanity's greatest habitat. The single biggest thing we can do for nature. So turn the pages of this book, gather your ideas, come up with a plan and let's get digging.

WHAT MAKES PONDS SO SPECIAL?

Of all of Britain's wetland plants and animals, two-thirds of species are found in ponds. Among the showpiece organisms in ponds are a host of threatened species that could not survive anywhere else. They include the pool frog, the natterjack toad and the great crested newt – each of these creatures is a specialist of ponds, particularly during its larval stage. Then there are the bats that feed along the water surface and the birds that feed at the pond edge. There are the hedgehogs and badgers that use ponds as places to drink, and the birds that use your pond to bathe. Your local pond is their local lifeline. And that's just the vertebrates. The plot thickens when we consider invertebrates.

For invertebrates, ponds are an entangled mess of energetic feeding relationships almost dizzying to those who study them. More than 4,000 species of freshwater invertebrate are known in the UK, and most of these are found in ponds. On many occasions I have undertaken surveys with schoolchildren on pretty standard-seeming wildlife ponds and uncovered more than 50 species, including dragonflies, water beetles and caddisflies. In some ponds, 100 species or more can be named in a single morning. With a microscope, the number of animals in a sample can quite easily triple so numerous are the unseen pond creatures, such as the rotifers and water fleas and hair-worms and hydra. Of the 4,000 freshwater invertebrates known, there are invertebrate species that specialise in surface-living, invertebrates that live at the pond edge, invertebrates that live on pond plants (as well as invertebrates that live *in* pond plants), invertebrates that live on the pond bottom and invertebrates that swim like sharks through the water column. So why? Why exactly are ponds so important for so many organisms?

The reason for this species-richness is partly because natural ponds contain a variety of microhabitats and niches that, to put it simply, allow lots of animals to do lots of different things all at the same time. Like rainforests and coral reefs, ponds are habitats that are three-dimensional in their structure. There are opportunities left and right for some animals to exploit, but also there are opportunities in the up and down space for other species. This three-dimensional structure provides more space and architecture for animals to fill. Ponds allow for niche ways of life, like hole-living, surface-skating, ambush predation, scramble predation, egg-laying, overwintering, sleeping, plus, well, a hundred different styles of mating. Ponds are busy communities. They are submerged natural cityscapes. One habitat, tens of thousands of opportunities. And that's why they are so special.

But ponds are not only useful for animals. They are also important refuges for a host of unusual and threatened plants. Ponds provide homes for most of Britain's 400 or so large wetland plant species and, of the threatened species of wetland plants, approximately half are known from ponds. And what of algae? Well, it probably won't surprise you to learn that there are many, many more species of algae out there. Like the micro-invertebrates, numerous unstudied or even unnamed species of these primitive life forms abound in ponds. Armed with a microscope and an eye for the tiny, many intellectual riches await new naturalists. For them, a garden pond is a training ground. A place to get an eye for the weird and wonderful.

How do animals colonise ponds?

Perhaps the most impressive thing about ponds is how quickly animals turn up to occupy them. In my first (quite urban) garden ponds, I observed countless fly species (all pollinators) laying their eggs in the water within only a matter of days. Within two weeks these ponds could be home to three different species of water beetles (not to mention backswimmers) all drawn in upon powerful flying wings to feed upon the fly larvae. Within 30 days, a frog might arrive. Within a year, newts.

What one realises upon digging a pond is that these animals, particularly invertebrates, are almost waiting for opportunities to colonise. They are flying overhead or moving along the floor through our gardens, looking for wetland opportunities. Without our pond digging, their search would have continued elsewhere. It may even have ended up in vain.

PREHISTORIC PONDS

Ponds have a rich history spanning millions of years. Thousands of fossils in museums around the world shed light on the waxing and waning of this precious habitat, and detail the lives of creatures that lived millions of years before the rise of modern humans. Here we explore how nature made ponds before we came along, and how we might make better ponds today by considering those ponds lost to us many millions of years ago.

ABOVE Many scientists, including Charles Darwin, were avid pond-dippers in their youth.

TREE-ROOT PONDS When a tree falls over, its underground root system is pulled from the ground, exposing a pit-like hole that is able to fill with water after subsequent rain. These tree-root ponds are not usually large, sometimes only a metre or two in diameter, but they quickly become home to numerous forest invertebrates and amphibians. Many tree-root ponds will fill up with leaf litter in the years that follow, but as long as other trees continue to fall in the vicinity, the survival prospects of freshwater plants and animals will be secure.

BELOW For millions of years, tree-root ponds have provided temporary homes for countless animals.

RIVER PONDS Many natural ponds form next to rivers. The classic is the oxbow lake, where the gradual erosion of a river edge cuts through and makes redundant a river or stream's meander, turning it into a long pond. But ponds form next to rivers in other ways. Sometimes a fallen tree or a collapsed bank, for instance, can dam up a section of river, slowing water flow and creating a temporary pond.

PINGO PONDS Pingo ponds are rare but can be very important locally. These circular ponds are infilled depressions in the ground left over from the last ice age 10,000 years ago. They represent the final resting places of giant lumps of ice whose weight caused a literal impression upon the ground as it melted and subsequently filled with water. Today, Norfolk is a stronghold for pingo ponds. Dozens remain in the Brecks, where they play an important role in helping local wildlife.

BEAVER PONDS Long before there were spades and mechanical diggers, there was the beaver. By damming up rivers and streams, beavers and their extinct relatives have been creating ponds for perhaps millions of years.

Beavers do this to deepen watercourses to such a depth that they can build a protective lodge in the middle. Within this floating structure they rear their young, protected from predators. Beavers also see out winter in safety hidden within these lodges. By doing this, beavers inadvertently engineer a habitat that can be used by hundreds if not thousands of other wetland species, including dragonflies, caddisflies and, of course, amphibians.

FOOTPRINT PONDS Even the footprints of some large megafauna can be enough to create tiny pools and ponds. Elephants walking through muddy grasslands, for instance, can create micro-ponds capable of holding 100l or more of water with each footstep. In 2014, scientists surveying just 30 of these water-filled footprints discovered 60 species, including beetles, worms and tadpoles. The same will almost certainly have been the case with the mammoths and their close relatives that once lived across much of what is now the UK. Large dinosaurs, including long-necked sauropods like *Diplodocus*, also created ponds with each of their footsteps. It may be that many prehistoric species, particularly frogs, depended upon some of these footprint ponds for survival.

So what can we learn from these prehistoric ponds, you may ask? The first and most important lesson to take on board is that many prehistoric ponds were transitory habitat features. In other words, they were ponds that didn't last for long. While oxbow lakes might last centuries before being infilled by leaves and sediments, tree-root ponds may only last a matter of years and footprint ponds a matter of weeks. A lost pond does not cripple an ecosystem, as long as there is another to replace it. This is the natural way with all ponds.

But there is another take-home message we get from examining these prehistoric ponds. It is simply that, on the whole, nature does not engineer deep holes. In fact, most prehistoric ponds probably only held water for a few months of the year and were quite shallow. They were seasonal habitat features, in other words. A dried-up pond is not lost for ever. Instead, it can be a habitat feature going through an ecological version of animal hibernation. Given time, life always returns. And often it comes back better than ever before.

PONDS AND CIVILISATION

Before humans drained the land for agriculture, perhaps a quarter of the British Isles was a wetland of some kind all year round. In fact, much of Britain was marshes, bogs and fens. This jars quite shockingly with the landscape we see today. So altered by humans have our natural habitats become that in the modern day only a few areas – such as the New Forest of Hampshire and the Caledonian pine forests of Scotland – give us a glimpse of what our 'green and pleasant land' was really once like. The idea that ponds were once rare is fallacy. In fact, natural ponds were probably quite common.

Undoubtedly, the most dramatic changes that have happened to our landscapes have taken place in the last 2,000 years, particularly with the invasion of the Romans and the subsequent rise in the UK's population. Rivers and streams became an obvious source of water for the advancing agriculture required to feed this growing population. By people straightening and deepening these waterways, Britain lost many of its meadows, its meandering trickles and micro-deltas that so many species depend naturally upon. Many ponds undoubtedly disappeared during these centuries.

However, though it is tempting to lament the loss of wetlands during this time, there were gains made. By understanding the value of freshwater to agriculture, hygiene and (in subsequent centuries) industry, humans began to store water in new ways for their own purposes. In recent centuries, they started building their own ponds to suit the needs of the people, and some of these ponds remain in use today.

BELOW Many village ponds were once vital top-up stations for horse-drawn carts.

VILLAGE PONDS In an age before running water, no settlement could last long without its own village pond. At first, people used these ponds as a drinking source for village animals or as stop-off points for livestock passing through to market. Later, they used them for bathing and cleaning and, later still, for more recreational pursuits, including boating and skating in particularly icy winters. During the Industrial Revolution village ponds served as an important source of water for steam-powered vehicles. In fact, many historic pubs are situated next to ponds specifically so that both human and vehicular thirsts could be quenched.

FARM PONDS To provide water for crops and thirsty plough horses, farm ponds become another freshwater resource added by humans to the British landscape. Some farm ponds were dug near to farm buildings to help power machinery whereas others were dug into adjoining corners of fields to provide for livestock. Some of these farm ponds remain today in the corners of fields. Many are still surrounded by willow trees first planted by farmers centuries ago keen for an easy and accessible source of firewood.

DEW PONDS As pastoral farming expanded during Victorian times, farmers needed to find new ways to provide water for livestock. By

ABOVE Despite the name, dewponds get their water from rain.

hollowing out a circular depression into chalk and lining it with clay or lime, farmers could create so-called 'dew ponds', which stayed wet all year. Dew ponds were once a particularly important source of water for livestock in various parts of southern England.

FISH PONDS As well as village ponds, many medieval settlements were also home to their own collections of fish ponds. Among the fish stocked for food were perch, pike, roach, bream and eels. These ponds, often dug within the grounds of monasteries and manors, became a kind of status symbol in the Middle Ages and each will have taken much energy and effort to maintain.

CLAY PONDS AND BRICK PITS

Today, some of the most important wetland sites in the UK are water-filled holes in the ground left after the excavation of clay for industrial purposes. During the Middle Ages, this mineral-rich clay was mostly extracted as a fertiliser (called marl) or for use in pottery. Later, as the industrial age led to a proliferation in urban house-building, clay provided a crucial component in brick-making. Today, many of these Victorian clay pits are important not only for nature but also as local fishing spots or recreational boating lakes.

MILL PONDS From late Saxon times onwards, early engineers began to toy with gravity-powered technology as a way to power the giant wheels of agricultural and industrial expansion. By creating vast mill ponds, for instance, water could be managed and specially channelled over water wheels, providing a source of plentiful (and free) energy. For the time, this was one of humanity's most impressive emerging technologies. Some mill ponds are still in use today.

THE BIRTH OF THE ORNAMENTAL POND

As modern civilisations grew, ornamental gardens gained greater and greater significance in the lives of royalty. Their showpiece gardens became a means for flaunting their wealth and botanical ingenuity to neighbours or visiting rivals.

Perhaps the first detailed record of water features in such ornamental gardens comes from the Hanging Garden at Nineveh, constructed by the Assyrian king Sennacherib (705–681 BC) for his palace, which is situated in what is now Mosul in northern Iraq. Apparently, to irrigate the gardens, specially created canals carried mountain water for 50km, including across an enormous aqueduct constructed from 2 million dressed stones, each lined with waterproof cement. Some historians consider these gardens to be the mistranslated location of the Hanging Gardens of Babylon (*pictured above*), one of the supposed Seven Wonders of the Ancient World. Undoubtedly, running water played a major part in gardens like this one. However, to what degree the water served an aesthetic purpose, rather than a purely practical one, is unknown.

In those early centuries, one thing that limited the use of water in ornamental gardens was surely the trouble it took to get it there in the first place. For this reason, water features like flowing streams and bubbling pools were restricted to ornamental gardens that happened to be next to existing streams and rivers. Only in more recent centuries would that change as technologies unlocked new mechanisms that allowed for the use of water in new and eye-catching ways. In the sixteenth century, for instance, the fruits of Greek philosophy, particularly advances in the human applications of hydraulics and pneumatics, allowed for novel sluices, gates and channels in ornamental running water features. Running water could do things like open mock temple doors, make folly fountains suddenly erupt and even make mechanical birds sing. Like a modern-day fad, many European palaces at the time invested in technologies like these to wow their audiences. Again, ornamental ponds were about status.

As decades and centuries passed, ponds continued to be the playthings of the rich. In Britain, those living in country houses and royal retreats used their ponds as a symbol of wealth and circumstance. In the eighteenth and nineteenth centuries, water features were specially designed and built in front of such properties, so that incoming visitors on horse and cart were immediately taken in by the spectacle. Landscape gardeners, such as Joseph Paxton (1803–1865), became celebrated for their trademark ponds, which really were great feats of engineering. Paxton's most famous ponds include the Canal Pond in front of Chatsworth House (made famous by the 2005 film adaptation of *Pride and Prejudice*) and the once state-of-the-art ponds (with added tidal simulations) in Crystal Palace Park, known

today for its illustrious Victorian representations of dinosaurs.

However, the age of ponds as playthings of the rich was not to last forever. Eventually, the industrial age brought with it a new invention: the lowly pond pump. This accessible technology allowed water to be recycled and redistributed in a pond without the need for a constantly refreshing source of freshwater. With this simple device, a new era of ornamental ponds would arise in the twentieth century and, as the cost of this technology lowered, so ponds (particularly ponds for fish) now became the playthings of the middle classes.

As new plastics continued to be trialled and produced in more recent decades, ponds could be lined much more easily too. They held water better and they became more foolproof. The prices of these plastic liners continued to drop, making ponds accessible to almost anyone with a garden. So dramatically did the British trend for garden ponds take hold that, by the end of the twentieth century, garden ponds constituted about 20 per cent of all shallow pond habitat in England and Wales. Without ever really being aware of their actions, an ornamental fashion for tranquil ponds and backyard fishkeeping had seen gardeners provide a lifeline for a host of creatures, including newts, frogs and dragonflies – creatures that were suffering substantial

declines in the wider countryside. Without the garden ponds created during the latter half of the twentieth century, the UK's wetland wildlife would be in a far sorrier state. And so, for once, a penchant for semi-urban and urban ponds provided a good-news story for fans of British wildlife. Together, we brought wetland nature back. But our work isn't done just yet, as you will discover.

THE STATE OF BRITAIN'S PONDS

Of course, we should shout from the rooftops about the positive impact that garden ponds have had on the UK's wildlife, but it also helps to be realistic. Though garden ponds have mitigated to a degree the pond losses that have occurred, the speed and scale of pond loss across the UK has been so dramatic that we have more work to do.

One well-worn statistic speaks volumes. Over a period of about 150 years, we have lost half a million ponds in the wider countryside, and with them literally millions of freshwater animals and plants have died off. In the twentieth century alone, this equates to a three-quarters loss of all British ponds, mostly because of drainage or infilling for agriculture. Even with the recent fashion for garden ponds, the creeping spread of our towns and cities has also taken its toll.

But it's not all doom and gloom. In the last 30 years, for instance, pond numbers appear to be on the rise. Though ponds are still being lost in the wider countryside, new ponds are helping to counter this loss. Today, ponds are becoming an important part of urban infrastructure, common to many new housing estates and retail parks, and

next to new roads and motorways. Planners understand the value of ponds in dealing with flooding, in providing places for people to walk their dogs and enjoy the water's aesthetic allure. Increasingly, as our understanding of mental health improves, we realise that communities appreciate urban spaces for nature, and that standing water plays an important role in this. And schools, too, have always understood the educational benefits of a hearty bit of pond dipping.

To a degree, losing some of our countryside ponds matters less if we replace them with something closer to population centres, something many more of us are capable of cherishing day to day. These newest ponds may help us do that. Judging by our efforts, urban ponds and other wetland areas could one day become a beating heart of the UK's wetland ecosystems. But there is a problem. It is a problem that has overtaken pond loss as a leading concern for conservationists and ecologists. That problem is pollution.

More than a decade ago, the Countryside Survey 2007 outlined a worrying trend occurring throughout our freshwaters. The survey revealed that 80 per cent of ponds in England and Wales were in a 'poor' or 'very poor' condition. Many of the pollutants discovered during these surveys were those associated with road run-off or nutrients associated with intensive arable land use. Acidification of waterways was another concern. The condition of ponds in Scotland and Northern Ireland may also be similarly affected, though perhaps to a lesser degree.

As the decades have passed, we have begun to make important steps towards replacing the ponds lost a century or more ago. New ponds are crucial in this endeavour. Clean, unpolluted new ponds even more so.

But there is another factor behind the pond losses we have seen in the British Isles, and it is one this book tries to remedy. It is simply that, with the best intentions, some ponds are managed the wrong way. There is a temptation to dig out and deepen every pond that is filling with leaves and vegetation, for instance, but actions such as this can ruin plant communities and kill off wildlife that has colonised specifically to live in such a habitat. Many ponds have natural lifespans. Some of the best ponds are temporary. The aim of this book is to underline this natural principle throughout, providing simple advice to readers so that they can make more informed decisions about pond management.

HIDDEN VALUES OF PONDS

By the beginning of the twenty-first century, ecologists, biologists and wildlife conservationists were clear about two things. First, that ponds were one of the most valuable and species-rich habitats in the UK. Second, that they had suffered a serious and sharp decline in the wider countryside. Something had to be done.

In 2009, the wildlife charity Freshwater Habitats Trust (formerly Pond Conservation) launched a partnership initiative called The Million Ponds Project. This project aimed to restore Britain's assemblages of lost ponds, taking us from 500,000 ponds to a pre-industrial target of 1 million ponds within a 50-year period. A key point in this ambitious target is that these new countryside ponds will be provided by clean water and will have been colonised naturally. But wildlife ponds can be about more than just wildlife; many have other values to people that we can easily overlook.

LEARNING OPPORTUNITIES

A simple net and white tray is all one needs to discover an enchanted world of zoological wonder, providing young and old with opportunities to see and explore educational topics including food chains, adaptations, anatomy and classification in a hands-on and very memorable way. Many modern scientists owe their careers to such early experiences with ponds at school or in field centres.

What might a twenty-second-century pond look like?

Ponds and their uses are likely to change again as the decades stack up. Here we consider for a moment how future generations might use ponds, in addition to conserving wildlife.

BATTERY COOLING As the renewable-energy revolution continues, many companies are examining how electricity from solar panels or wind generators could be stored locally and reused by homeowners in special batteries. Yet many batteries can get hot when in constant use and so will need cooling. Perhaps in the future many homes will use renewable water sources to undertake this cooling. Ponds could have a new purpose in such a world.

CARBON CAPTURE By being so productive, ponds remove carbon from the atmosphere at about 20 to 50 times the rate at which trees capture carbon. In fact, ponds take up carbon at a higher rate than many other wetland habitat features, including lakes. With this in mind, it may be that in the future world economies look to mitigate the impact of their fossil-fuel technologies by pushing ponds right to the very forefront of their environmental initiatives.

BIOFILTRATION In many parts of the world, people use specially constructed ponds as a way to capture and degrade pollutants found in water. In these ponds (called waste stabilisation ponds) water passes through a number of pools, in which known pollutants are trapped or broken down. In the future, could homes use their own biofiltration systems to treat and reuse waste water, thereby saving water? Such a system would dramatically reduce the amount of water that we, as a society, consume.

PONDS FOR SCIENCE More and more, ponds are becoming accessible habitats that can provide scientists with important data. Such ponds can be evaluated and included in many citizen-science projects, helping scientists understand the role that garden ponds play locally. Take frogs and toads, for instance. The British Trust for Ornithology includes amphibians like these in their comprehensive garden surveys, helping scientists understand more about which gardens are suitable for amphibians and how we might encourage their spread further into urban areas. Another example is the Zoological Society of London, which runs a project that keeps track of unusual die-offs caused by non-native frog diseases. Projects like these depend totally on the public acting as the eyes and the ears of the nation, looking out for unusual incidences of diseases in the wild (see page 125).

WELL-BEING AND GOOD TIMES

In recent years, social scientists are beginning to put value on the importance of ponds and other natural settings for the range of mental health benefits they can bestow upon people. For many pond owners, garden wildlife ponds are a place to unwind and to consider the world without the trappings of modern technology. More and more, we are coming to understand the physical and mental benefits of time spent near outdoor ponds. They are places to de-stress. To meditate. To think clearly. To breathe.

RIGHT Ponds allow us a quick and easy opportunity to plug into nature, offering a host of mental health benefits.

A VERY SOCIAL MEDIUM Lastly, there are the other pond owners for whom ponds have become a place to marry nature with modern technology marvellously through social media. Some of my favourite social media accounts detail the daily goings-on of their ponds, providing a soap opera-like account of the drama. Some use microscopy to explore ponds in new and unexpected ways. Others use social media like a nineteenth-century nature diary, mixing science with literature, poetry or simple commentary on personal growth. Ponds can do strange and important things to people, as you will hopefully come to discover.

THE IDEAS STAGE

AT THIS POINT, you can consider what size and design of pond might suit your outdoor space. Remember that to ensure the maximum number of visitors to your pond, you will have to mimic the natural ponds to which they will have adapted over millions of years. Take your time over this stage. The choices you make now will have a dramatic impact on the wildlife credentials of your pond further down the line.

The main principle to keep in mind is that your pond needs complexity. By creating a variety of depths, shores and shelves in your pond you can make it attractive to a greater range of plants. The more plants you attract, the more nooks and crannies in which animals can hide, and the more feeding, hunting and egg-laying opportunities there will be for other organisms such as dragonflies, bugs and beetles. These animals bring with them their predators – the wow-inducing newts, frogs, birds, bats and other mammals. To attract inspirational creatures like these, always remember that complexity is king.

CONSIDERING CREATURES

All ponds have value but the best ponds are built with an understanding of the needs of their potential residents. Smaller invertebrates, for instance, are drawn into patches of water in their search for food or because they are looking for specialist plants upon which to lay eggs. Some animals visit because they are seeking protection from predators. Others are seeking something a little different – a quiet place to see out the winter months, perhaps, or a place to hunt tadpole prey. The best wildlife ponds have something for everyone. They are filled with niches, in other words. They abound with opportunity. In this section, I introduce some general points worth considering as you begin thinking about your future wildlife pond.

ABOVE Marking out your pond can help you get a feel for how much light it might get.

BE BOLD Ponds needn't be circular or symmetrical. They can be made far more complex and niche-filled by the addition of miniature bays and banks, much like the ponds that nature builds. These increase the complexity of the pond's shoreline, enhancing the chances of multiple species occupying the same area all at once. Likewise, the pond floor can be made more interesting. Slightly deeper sections (up to 50cm) with deep-water planting add a new layer of complexity to the pond habitat. At every stage in your pond design, ask yourself the question: 'How can I make this more interesting?'

ABOVE Common frogs may be one of the first charismatic vertebrates to visit your pond.

ABOVE Adding cover near your pond will help keep your animals safe from cats.

BE ACCESSIBLE Ideally you will want animals and plants to colonise your pond naturally, so the question is … will they be able to get into your garden? If you are particularly interested in attracting frogs, for instance, you will need to consider your garden walls. Are there gaps through which frogs might move? Is there long grass or thick foliage through which they might travel to keep cool and stay hidden from predators? Are there ponds or ditches within 300 or so metres? If your answer to questions like these is a resolute 'no', then what can you do to change this? Solutions include cutting little holes underneath your fences for wildlife to travel through or encouraging neighbours to add ponds too, creating an urban mosaic of 'stepping-stone' freshwaters like those that exist in the wider countryside. We explore the accessibility of wildlife ponds in more detail on page 60.

PONDER YOUR PETS In many urban locations, predatory pets are likely to be an issue for your pond. The presence of large, noisy dogs or stealthy feline predators, for instance, means that animals like frogs may be tempted to avoid your pond. You can encourage wetland animals to stick around by offering them pet-free 'panic rooms' that surround the pond. The gaps within log piles and/or the tangled labyrinths of plant roots, or underneath piles of stones, can be crucial places to seek temporary reprieve from cats, for instance. Through simple additions like these, you can make a pond where cats, dogs, frogs and other smaller pond animals are seemingly happy to co-exist.

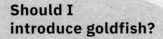

Should I introduce goldfish?

Though undoubtedly beautiful, many ornamental fish are specialised hunters of aquatic invertebrates, so in all but the largest ponds their presence can dramatically affect wildlife communities. The Freshwater Habitats Trust reports that 50 per cent of all pond species will not thrive alongside fish, including ornamental fish such as goldfish. So introducing fish might hinder rather than help your pond's chances of becoming an important freshwater wildlife spot. If you love ornamental fish and you have the space, could you consider digging two ponds – one for the wildlife, one for the fish?

WHERE TO PLACE A POND

A well-sited pond might more than double the number of species that visit, so you should consider carefully where in your garden or backyard your pond could go. In this section, we explore some of the main things to remember when choosing a possible place for your pond.

LOCATION, LOCATION, LOCATION

Putting a pond in full sun maximises the energy that plants and algae can capture, invigorating your pond's food web. But shallow ponds will get warmer in summer, resulting in a pond that may dry up too quickly. Some animals, including tadpoles, may fare badly in such conditions. For this reason, many pond owners prefer a partially shaded location for their ponds rather than putting one in full sun. This means that animals can migrate within the boundaries of the pond to seek out the spots that best suit them.

CONSIDERING TREES Trees can offer ponds useful shelter from the sun, but remember that, in many cases, their fallen leaves will fill your pond in autumn. Though small amounts of leaves are undoubtedly a good thing for ponds, since an army of detritovores (including small crustaceans) rely on them for food, smaller ponds are likely

to fill up more quickly than they otherwise might and may end up being choked out completely within four or five years. Tree roots can also be a problem. Fast-growing roots can misshape the pond liner and lead to punctures.

FUTURE-PROOFING YOUR POND

Always plan a pond with the future pond owner in mind. For instance, will you have the time and inclination to rake leaves away from the edge of the pond each autumn? If the answer is no, avoid planting trees nearby, or siting your pond too close to trees. Do you think you might consider a dog at some point? If so, soft edges might quickly be ruined by their scampering claws. And might there be children or grandchildren playing near the pond one day? If the answer is yes, you may prefer a different location in the garden. Questions like these aren't designed to put you off digging a pond. Rather, they help reduce the likelihood of having to re-site the pond in future years – not an easy task!

KEEP IT LEGAL Most garden ponds will not require formal planning permission but it is good practice to phone your local planning authority to check, particularly if you are planning a larger pond. Large ponds may require the use of a digger, which can be noisy for a day or two. It can be common courtesy to let neighbours know what

to expect by way of noise or, for instance, limited access to shared driveways.

If you hope to dig a pond on any land that is not privately owned, you will need formal planning permission. Likewise, if your pond will be filled by a river with run-off water that then returns back to the river, you will need to confer with your country's statutory nature conservation agency (see page 155), which may require that you possess a special licence. If you are considering moving or cutting back trees you will need to check whether or not each tree has a Tree Preservation Order. Rushing over precautions like these can be very costly by way of penalties, so take your time to get things right first time.

KEEP IT CLEAN The most important ingredient for your pond is clean water. The cleanest water comes from the sky, and in every case this is the unpolluted water that you should use to fill up and refresh your pond. Though the temptation will be to top up your pond with tap water – especially in dry summer months – try if possible to hold out for rain. Tap water has additives, including nitrates, which can accelerate the growth of problematic algae and pondweeds further down the line.

HOLDING WATER

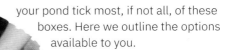

Because urban soils are so well drained, you are likely to need a liner of some sort to provide an impermeable barrier to stop your water seeping away. This barrier needs to be weather-resistant and not likely to be broken down by bacteria. It also needs to be resistant to ultraviolet damage from the sun and it should be strong enough not to split should a sharp twig or an animal's claw ever snag against its surface. If within or nearby a public place, this impermeable barrier may also need to be resistant to vandalism. Only a few options for lining your pond tick most, if not all, of these boxes. Here we outline the options available to you.

FLEXIBLE SYNTHETIC LINERS

Most garden and aquatic centres stock a range of flexible plastic products, which offer a number of advantages, particularly in terms of price. Each of these flexible liners is durable and easy to cut to size, and most can be very long-lasting in undisturbed ponds. On the downside, punctures in flexible synthetic liners can be hard to repair and the liner on its own is not a natural substrate upon which vegetation can grow. For this reason, you may require a layer of subsoil.

PRE-CAST PLASTIC LINERS

Many different kinds of sturdy pre-formed pond liners (left) are sold in garden and aquatic centres, and each may last 20 years or more. Often designed for fishkeeping, these ponds can have high sides, which make them hard for visiting amphibians (and hedgehogs) to get in and out. This can be remedied with the addition of well-placed rocks, specially planted foliage or log piles on shore areas that offer something by way of a makeshift ladder or stairway.

PUDDLED CLAY

The traditional way to provide a waterproof lining for ponds was to cover an excavated surface with lumps of clay cut like paving slabs, each carefully slotted against one another as tightly as possible. When watered, the joins would then be trampled (in the old days using livestock) to compact (or 'puddle') the clay into a complete layer of uniform thickness. Today, rubber-tracked vehicles can provide a useful alternative to those without access to sheep and cattle.

BENTONITE This naturally occurring non-toxic clay swells to 10 or even 20 times its dry volume when wetted. Often it comes as a powder-like substance that can be difficult to lay evenly upon newly dug ponds. For this reason, many pond owners only use bentonite as a way to repair pond leaks or seal the edges in previously clay-lined ponds.

GEOTEXTILE CLAY LINER For garden ponds both large and small, geosynthetic clay liners (GCLs) offer the naturalistic perk of having a clay-lined pond while being fairly easy to install. Strips consist of a layer of bentonite clay, sandwiched between flexible geotextile.

CONCRETE For the specialist only. Often concrete linings may be poured into geometric shapes (mostly rectangular), which can make them look less natural. If subsidence should occur, they are prone to cracking, which makes them very time-consuming to repair.

THE PROS AND CONS OF DIFFERENT POND LINERS

type of liner	pros	cons
Puddled clay	• Ready-to-go substrate for plants and animals • Easy to repair	• Incredibly labour-intensive • Will possibly not work first time • Edges will crack during dry periods
Bentonite	• Can be used as a 'booster' for naturally clay-rich soils • Easy to apply and to repair	• Not suitable for smaller ponds • Machinery (rotovator) required
Geotextile clay liner	• Adaptable product, with strips able to be sealed easily on site • Self-healing • Forms natural substrate for plants and animals	• Rolls are heavy • May require machinery to lift and apply strips on site
Flexible synthetic liners (including EPDM liners, polythene, LDPE and PVC liners)	• Very flexible • Easy to cut and apply • Small/medium ponds can be installed by non-specialists	• Limited lifespan • Requires underlay • Often liner tears will require a completely new replacement liner • Difficult to recycle
Pre-cast plastic liner	• Easy to install • Difficult to puncture • Very long-lasting, sometimes 20 years or more	• Requires carefully prepared compacted hole to avoid subsidence • Vertical edges will need to be mitigated. • Difficult to recycle
Concrete	• Incredibly hard-wearing	• Costly to install • Steep edges and large size make these ponds less suitable for wildlife • Sometimes fracture in cold weather or through subsidence • Very difficult to remove

PROFILING YOUR POND

A key part of your decision-making process should be how you can maximise the opportunities for wildlife in your potential pond. You can do this by considering your pond in three dimensions, thinking about the water surface and the water substrate, and the water column that connects these two realms.

PLAN FOR DROUGHT You should prepare your pond to have large shallow sections that will become your 'drawdown zone' in long dry summers. This part of the pond forms one of the most important areas for developing invertebrates and amphibians. Two or three times as many species may live here than in deeper water. Many dragonflies choose to lay their eggs in the drawdown zone, for instance, safe from pond predators. Other animals like pirate wolf spiders frequent these areas, as well as shore bugs and ground beetles. Water shrews and wading

BELOW The gaps between roots can provide superb hiding spots for amphibians. Believe it or not, there are eight froglets and toadlets pictured here!

birds may also stop by, searching for partially submerged prey among the shallow vegetation.

For plants the drawdown zone may be even more important. Many seeds of marginal wetland plants, for instance, depend on exposure to air to begin their growth. With their germination comes even more niche partitioning, thereby guaranteeing the pond a wealth of different animals in future years.

ROOT GARDENS

Some trees (including willow) and pond plants (such as yellow flag iris (*Iris pseudacorus*) and bulrushes) create vast mats of labyrinth-like roots, which many animals seek shelter within, particularly if disturbed. In fact, some resemble mini-mangroves, and may be comparatively as rich biologically speaking. Plan for these root structures when considering your pond's plants.

WETLAND MARGINALS

Overhanging or semi-floating herbs and grasses can become incredibly fruitful places for many invertebrates to find food and seek shelter within. In fact, a simple sweep with a pond net in these locations can sometimes expose 10 or 20 different invertebrate species.

POND LAGOONS

Shallow regions that level off to become their own shallow puddles within the drawdown zone can be especially fruitful places for pond animals, particularly smaller water beetles and amphibian larvae eager to seek a predator-free hideaway from, for example, fish and voracious diving beetle larvae. Plan for these in your final designs.

DEEP WATERS

If the pond edge is like a miniature coral reef, the deeper waters of the pond are more akin to the open sea, offering far less by way of opportunities and, therefore, less species diversity. For this reason, most wildlife ponds need not be more than 50cm deep. This is not to say that deeper ponds offer nothing for wildlife, however. Frogs may choose to spend winter in a dormant state in such ponds, and bats like to hunt over areas of deeper water that lack vegetation. Diving ducks, too, may visit larger ponds with deep sections (not more than 50cm deep) in search of submerged offerings.

SHAPING YOUR POND

Viewed from above, you should imagine your pond as having two shapes: one in winter (when the water is high) and one in summer (when the water is low). Your pond plan will need to cater for both. Here are some general pond shapes, each with their advantages and disadvantages considered.

CIRCLES While a singular circle-shaped pond can offer a certain aesthetic appeal, unless planted wisely, their banks offer little by way of variation. To add great prospects for wildlife you could consider including 'satellite' ponds of various sizes around the edge, each with different depths. A quick sweep with a net or a bit of nighttime observation with a torch should

OPPOSITE/BELOW **1** An example of a non-complex shape. **2** Amphibians may struggle to get in and out of a pond with an edge like this. **3** A multitude of edge habitats on display. **4** A gravel shoreline is aesthetic but can be a lifeless desert in summer. **5** A classic kidney-shaped pond.

show different species utilising each of your connected puddles and ponds.

THE KIDNEY BEAN This is the classic garden wildlife pond shape, recommended by many because it mimics more closely the ponds that nature builds. The kidney-bean shape increases the perimeter of the pond edges, allowing more space for edge-specialists, like some water beetles, bugs and frog tadpoles. A curved outer bank can expose more of the pond to the sun, ensuring it absorbs more energy than it might otherwise have managed.

THE WAVY-LINED POND Having a more complex edge to your pond, which allows for deep areas and sloping shores, can prove very attractive for wildlife. Flanked by harbour-like enclaves, certain species will gather to seek shelter from open-water prey. In drier months, these may turn into a network of tiny pools. However, it should be said that creating complex pond shapes such as wavy lines using a flexible plastic liner can prove very tricky.

5

TONICS, PUMPS AND FILTERS

Great swathes of filtration equipment, costly chemicals and pond tonics exist to make your pond apparently clear and beautiful, but the good news is that you are unlikely to need a single one of them if you intend to build a wildlife pond as nature intended. This is because most of these products are mainly for ornamental fishkeepers who, understandably, can be very keen to keep their water clean-looking in conditions where many fish live closely together, feeding and excreting large amounts in a relatively closed system.

If you don't have ornamental fish in your wildlife pond, you'll be pleased to know that nature provides its own filtration and oxygenation systems, which are outlined in this section.

NATURE'S WATER FILTERS It is natural for many ponds to go cloudy during winter as algae multiply while snails and water fleas lay dormant. As spring approaches, these algae-hunters

ABOVE **1** *Potamogeton crispus*, the curled pondweed – a classic oxygenator. **2** Pond snails excel at hoovering up algae. **3** Water flea blooms can make ponds crystal clear in a matter of days.

awaken, and in subsequent weeks multiply to make the most of the abundant food source. Water fleas are especially important to wildlife ponds. By birthing smaller clones of themselves again and again and again, they can increase their populations exponentially, meaning that algal problems in some ponds can sometimes seemingly disappear in as little as 24 hours as hundreds of thousands of hungry mouths go to work.

OXYGENATORS

The term 'oxygenators' is often used when describing some species of submerged pond plants, yet their importance in creating a healthy pond may have been slightly over-egged in the past. While it's true that many aquatic plants do provide a net oxygen gain, it is important to remember that ponds are naturally lower in oxygen content than many rivers and streams, and that the animals that live there have had many millions of years to adapt to such a low-oxygen habitat. Some animals, such as some pond snails, bloodworms and water hog-lice (*Asellus aquaticus*), have impressive tolerance for low-oxygen environments.

SURFACE SURVIVAL

The surface waters are the lungs of your pond. It is through this membrane that oxygen and other gases pass in and out from the atmosphere into your pond and vice versa. The ratio of surface area to water volume is really important when considering ponds. A deep pond with a limited surface area is like a big-bodied creature with a tiny set of lungs – a recipe for disaster. For this reason, shallow ponds with a large surface area are often among the most healthy and diverse pond habitats.

BELOW Sun-chasing flowers of common water crowfoot (*Ranunculus aqautilis*).

CONSIDERING PATIO PONDS

Budget: £100 to £2,000

Not everyone has a beautiful lawn or flowerbed into which can be dug a wonderful wildlife pond. Many flats or smaller houses only have space for a tiny outdoor patch, a concrete patio and not much more. Raised ponds are the best option in gardens like these. Though smaller than many ponds, they can be made to be very attractive with a little careful consideration.

WHAT IS A PATIO POND?

A typical patio pond is a raised flower bed within which is secured a moulded plastic pond liner. Being raised, they are easier for passers-by to study and observe, which can make them aesthetically pleasing. The downside is that some animals will need a helpful staircase to reach the upper tier upon which the pond is sited. Thankfully, most larger garden animals (including hedgehogs) are effective climbers. Rest assured; if they want the water enough, they will find a way.

WHO MIGHT PATIO PONDS SUIT?

Those with mobility problems. Those with small children or dogs. Those with patios, small gardens or outdoor decking that means digging a hole is impossible. Those seeking an easily accessible pond for educational use.

RIGHT Even the lower tier of an ornamental fountain properly planted up can provide a temporary habitat for freshwater animals (including frogs) as long as they have access via a handy pile of logs or otherwise.

WHAT ANIMALS MIGHT THEY ATTRACT?

A whole range of pollinating flies will make their journey to the pond within the early weeks, and they are likely to be followed by other creatures brought in by bathing birds and pond plants in the days and weeks after. Ramshorn snails and great pond snails (*Lymnaea stagnalis*) become one of the first grazers of the ripening algae. Then come the flatworms. Then come the crustaceans, eager to feast upon detritus washing into the pond. Principal among them is the pillbug-like water-hog louse (*Asellus aquaticus*), a specialist detritivore. Other crustaceans include water fleas. The plastic walls of ponds like this can sometimes be home to freshwater jellyfish-like animals called hydra. These strange creatures can clone themselves to create vast colonies. Patio ponds can become important summer refuges for frogs and newts, provided there is plenty of cover from predators.

Newts, particularly, may stick around for weeks, hunting like ghostly sharks the tiny bloodworms and other smaller creatures in the water.

PATIO POND FEATURES

Dizzy heights Generally speaking, you should aim for your pre-moulded pond to have a maximum depth of about 30–50cm. You can place this on top of a raised bed 1m off the ground. This ensures you have good access to your pond but also that larger animals can get in and out of the pond as required.

Stairway to heaven To allow animals such as amphibians and garden mammals to reach your pond you will need a staircase. Logs arranged into a neat pile (with no more than a 45 per cent gradient) can do the job very well, while also providing supplementary food by way of spiders and woodlice.

BELOW Water-beetles are occasional visitors to small ponds.

ABOVE Ramshorn snails sometimes thrive in small ponds.

Underfloor eating You can use the space underneath your pond to encourage the invertebrates that may become a vital food source for amphibians, including young froglets in summer. Rest your pond upon pallets and make a 'bee hotel' lined with bamboo to provide nesting areas for solitary bees. You can line holes with wall flowers to encourage pollinators to your pond. Habitat features like these encourage larger pond creatures like newts to stick around too.

Escape houses Patio ponds may well be in more urban places where predatory cats abound. Your pond will need an area of overhanging paddle-like stones, which frogs can hide beneath for safety. Other rocks will also be needed to ensure that pond animals can easily get in and out of the water. These rocks can double up as useful basking sites for those frogs eager for an extra bit of warmth in early spring.

CONSIDERING CLASSIC GARDEN WILDLIFE PONDS

Budget: £300 to £1,000

This is the classic wildlife pond design that many nature conservationists heralded and encouraged garden owners to dig in the latter twentieth century. Relatively safe and fairly easy to manage, a pond of this size has the potential to draw in a range of wetland creatures without costing the earth.

WHAT IS A CLASSIC GARDEN WILDLIFE POND?
Dug into the ground and lined with a flexible plastic liner, a pond like this might last 15 years or more. No deeper than 50cm, this pond will probably stay wet all year, though the water level will drop during long dry spells in summer. Shallow edges on each side

ensure that creatures, including hedgehogs and froglets, can easily enter and exit the pond. Many classic wildlife books and magazines recommend a kidney-bean shape, but because you dig and line the pond, there is flexibility to make a shape that better suits your garden if required.

WHO MIGHT CLASSIC GARDEN WILDLIFE PONDS SUIT?
Anyone with a garden that receives adequate sunlight. Anyone with enough time to do a spot of gardening and general sprucing-up every now and then. Because garden wildlife ponds are relatively shallow, siting such a pond underneath or near trees that shed their leaves is not recommended.

WHAT ANIMALS MIGHT THEY ATTRACT?
At this size, ponds become more interesting to water beetles and other aquatic invertebrates. You may have visits from damselflies, eager to explore new potential sites

in which to lay eggs. You will see backswimmers and a range of water beetles, some 3 or 4cm in body length. You are likely to notice pond skaters drifting along the surface. If there are newts in the area, they will quite happily set up shop for a summer, taking advantage of the hunting opportunities you have provided. In many parts of the UK, a pond like this one should encourage local frogs to spawn, sometimes in impressive numbers and within only a year or two.

LEFT Garden ponds can provide temporary homes for hundreds of water beetles.

BELOW A smooth newt uses its large eyes to hunt for pond invertebrates.

CLASSIC WILDLIFE POND FEATURES

Soft edges Soft edges rather than hard slabs mean that pond edges keep cooler in the summer months. This means that escaping froglets won't get cooked as they leave. The edges can be made of unrolled turf or planted areas held in place with coir rolls. Remember that soft edges can increase the rate at which water evaporates from the pond.

Southern-facing basking areas Rocks strategically placed on the sunny side of the pond can provide important places for young tadpoles and other cold-blooded creatures to bask under the water. Sometimes tadpoles gather in great numbers upon such rocks, often very near the surface. This is partly because they offer warmth from the early-spring sunshine but also because this is where the algae grow best.

Dense planting If your pond is at floor level it will need an area of dense vegetation at the furthest end from the sun to offer animals protection from cats and other ground predators. This planted area can be aesthetically quite pleasing, making your pond more three-dimensional and adding a bit of perspective to your pond design.

Predator perches

Damselflies (and sometimes dragonflies) frequent many garden ponds. Occasional perches offer them a stop-off point from which to look for possible prey, particularly non-biting midges or mosquitoes emerging from the water.

Floating-leaf pond plant area

The central part of a pond of this size should be open enough to allow for floating-leaf plants such as water lilies or water soldiers. These pond plants provide important sites upon which invertebrates may lay eggs and they also provide important shade for some pond creatures, like newts, in the summer months. For more on pond plants see Chapter 4.

Newt egg-laying sites Within a year or two, it is possible that your pond will be home to newts, particularly the smooth newt (*Lissotriton vulgaris*). Smooth newts lay their eggs one by one on pond plants. After a few weeks their fish-like larvae can be spotted darting through the pond. Suitable plants to attract egg-laying newts are described on page 91.

CONSIDERING LARGER GARDEN WILDLIFE PONDS

Budget: £3,000 to £20,000

If you have the space, the time and the money, larger wildlife garden ponds can be hugely valuable for local wildlife, attracting all sorts of showpiece species. In some cases, ponds such as these can become the ecological hub of local neighbourhoods, providing important nursery grounds for a host of animals that will roam the local landscape in their adult life stages.

WHAT IS A LARGER GARDEN WILDLIFE POND?

For the purposes of this book, a large wildlife pond is defined as a pond more than 10m x 10m in size. These are carefully considered ponds that offer a host of ecological niches that animals will colonise, often within a matter of months. Ponds like these may have sections of deep water in places, with sloping shallows, drawdown lagoons and areas of floating-leaf plants within which many invertebrate species will cluster.

WHO MIGHT LARGER GARDEN WILDLIFE PONDS SUIT?

Large ponds like this suit gardens with lots of space. They may also suit nature reserves or other public or private spaces such as schools or businesses looking to provide meaningful places for studying, observing and otherwise getting face to face with nature. Ponds like these may be very costly to install and will require the use of diggers. However, the rewards by way of colonising animals and plants, and the impact that ponds like this can have on neighbourhood nature make large ponds more than worth it.

WHAT ANIMALS MIGHT THEY ATTRACT?

Ponds of this size are attractive to a host of larger and more widespread wetland animals. For instance, rather than just attracting common frogs (*Rana temporaria*) and smooth newts, large ponds may attract local common toads (*Bufo bufo*) eager to invest in new breeding grounds. Sometimes, in parts of Britain, great crested newts may also make a surprise appearance. A host of pond predators may also appear in time. These include grass snakes (*Natrix helvetica*), grey herons (*Ardea cinerea*) and even the surface-hawking Daubenton's bat (*Myotis daubentonii*).

FEATURES OF LARGE WILDLIFE PONDS

Islands Small islands can be important predator-free places for nesting birds, including mallards (*Anas platyrhynchos*) and moorhens (*Gallinula chloropus*). They also make peaceful basking places for grass snakes. Some pond owners even add piles of bark and woodchip specifically to provide special nesting places for them.

Pond platforms Well-considered wooden platforms can be excellent places from which to access your pond with nets and other sampling equipment. Alternatively, they can be a peaceful place to sit and dip one's feet into the water. Platforms need to be carefully installed however, particularly if the pond is built using a liner. They will also need regular checks to ensure that foundations are not becoming rotten.

Pond edges Planted coir rolls can be a cost-effective way to plant larger ponds. These firm rolls hold the banks of the pond in place, ensuring that they don't collapse into the water. Remember to include regular access points so that you can get to your pond while surveying with a net or torch.

Open areas of water Not all animals like to hide among the shadows. Some animals, including the tadpoles of the common toad, like to move through open water in vast shoals. The larvae of great crested newt also like such places.

Deep areas Many larger ponds dry up over the summer months, so having three or four deeper areas in the pond can create a collection of lagoon-like pools at the bottom of the pond, which remain after the rest of the pond dries up. These pools may become a lifeline to the pond's aquatic life.

Hibernacula To keep amphibians near your pond throughout the year, you could consider hibernacula. These are special areas in which frogs, newts and toads (as well as some reptiles) may see out the winter in safety. For more on this habitat feature, see Chapter 2.

Supplementary ponds If you have the space, add another pond. These smaller supplementary ponds may attract different species of animals compared with the large pond, enhancing the total biodiversity of the local area. Supplementary ponds also offer opportunities for different aquatic plant species to blossom. If there is space near to your large pond, the more extra ponds you can add the better.

PREPARING YOUR POND

- -

IN THIS CHAPTER, we offer guidance and a step-by-step guide to building three types of ponds.

1 The first and easiest to install is a small patio pond, suitable for those with limited space or who live within highly urban locations.

2 The second is a classic garden wildlife pond, not more than 10m in diameter at its widest point.

3 The third pond style is for those seeking to create larger wildlife ponds that exceed 10m in diameter.

After you have lined and filled your pond, you'll be ready to add some features including basking areas, bird boxes, rockeries and compost heaps to ensure visiting animals spend more than just a few brief moments in and around your pond. Details of these accessory features are also included in this section.

In these early stages of your pond-planning, remember to factor in complexity. The finest ponds are often more than just a circle-shaped hole in the ground. The more gaps and hiding places there are in and around the pond, the more organisms will visit. In this way, ponds really can end up like coral reefs, filled with niches in which a diverse array of animals are busily getting on with their daily jobs.

PREPARING YOUR PATIO POND

For those who live in highly urban areas that may have limited outdoor space, a patio pond allows you to enjoy the magic of freshwater life. This includes some schools or those that live in downstairs flats or houses without proper gardens. Those with mobility issues can also create and tend to patio ponds. Let's start by making a simple raised patio pond by using railway sleepers.

Assess how much space you have to work with. Remember that you may need access to all sides of the pond, and that one edge of the pond will need some form of cover (provided through potted plants or a flowerbed) to allow animals respite during warm spells.

Purchase your wooden railway sleepers. Though costly, railway sleepers are strong and sturdy and resistant to weathering. You may need these to be cut down to size, a service that your garden centre may provide for free. You will need at least three tiers of railway sleepers.

1 To prepare your base for the railway sleepers, you may have to move some slabs.
2 Use concrete and sand to secure the bottom layer of railway sleepers and use a spirit level to ensure they are level. Double-check your levels. Mistakes now can impact how well the final pond comes out.

Add a layer of old carpet or cardboard into the pond area on which your liner will be laid. Use sand to create a smooth profile, creating shallower regions such as a small bay. Remove any sharp sticks or objects that might lead to punctures.

3 Add one (or more) tiers to your pond, using screws to fix the sleepers together.

4 Before adding your final tier of railway sleepers you will need to add your liner. Bricks can help hold your liner while you smooth it into place.

5 Begin filling your pond using rainwater if possible from a nearby water butt. As a last resort, tap water from a hose is another option though this may lead to algae problems further down the line. At this point 1cm of compacted soil can be added to the pond bottom, to provide a substrate on which plants can grow.

Whilst filling, the weight of the water will make the liner tighten. Remove the bricks to stop them falling into the pond and puncturing your new liner.

6 Once topped up and level, you are ready to add your final tier of railway sleepers. Cut your liner down to size with a pair of scissors. Your final tier of railway sleeper can be placed and secured over the lip of your liner, keeping it neat and hidden.

Finish filling up your pond before adding pond plants (if you do not wish to wait for plants to colonise naturally) and the gravel for your bay. Remember that you will need to wash the gravel before adding it to your pond.

Add a stairway so that larger pond animals, such as frogs and newts, can get in and out of the pond. A log pile along one edge, arranged into a stairway, can help here (pictured below). Likewise, you will need to create a way out of the pond should animals such as hedgehogs fall in. Add more pebbles or specially placed logs in your bay area to provide safe exit points.

DIGGING THE CLASSIC GARDEN WILDLIFE POND

Here is your 10-step guide to building a pond up to 10m in length. In this example we use a pond lined with a flexible plastic liner.

1 MARK OUT THE SHAPE OF YOUR POND To give you guidance while digging you will need to have an outline of your pond's shape sprayed onto the ground. For circular ponds, you can do this by placing a stake in the centre of your pond location and attaching string measured to the correct radius of your pond. Hold the string taut and walk around the stake, spraying the ground with a non-toxic biodegradable spray as you walk. You should end up with a perfect circle. If your pond design is not a circle shape, you can draw directly on the grass should you have the artistic prowess. Just remember to remeasure your pond's sprayed outline a few times to ensure that you will

definitely have enough plastic liner.

Once you think you're happy with your pond's shape, take a moment to consider it from all angles. Check it from a raised point like an upstairs window. Is it more overshadowed than you thought? Is it the right shape? Will there be adequate space for a walkway around each side? It's always worth asking these questions now, before it's too late.

2 GET DIGGING For a pond of this sort, you may prefer to do the digging yourself rather than employing a mechanical digger for the day. If so, the first thing to do with your spade is to carefully remove the turf from the surface. Cut it into long lengths and roll these up to keep them damp. Put them to one side near the pond because you'll need them later. Once you have a pond-shaped space devoid of grass it's time to dig through the topsoil. The simplest way to do this is to cut square-shaped plugs of soil from the ground, working across the soil in rows. You will have time to shape the pond later.

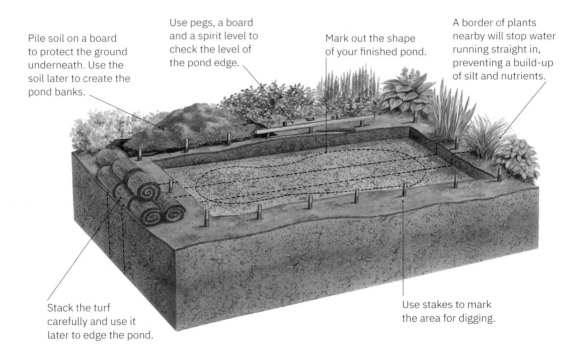

Pile soil on a board to protect the ground underneath. Use the soil later to create the pond banks.

Use pegs, a board and a spirit level to check the level of the pond edge.

Mark out the shape of your finished pond.

A border of plants nearby will stop water running straight in, preventing a build-up of silt and nutrients.

Stack the turf carefully and use it later to edge the pond.

Use stakes to mark the area for digging.

How much pond liner will I need?

To work out how much pond liner you'll need, use this rough-and-ready calculation. The total width of liner needed is the maximum width of the pond plus twice the depth of the pond. Likewise, the length of liner needed will be the maximum length of the pond plus twice the depth of the pond.

For a pond that will have a maximum depth of 0.5m and that is 3m wide by 5m long, the liner would need to be: 3m + (2 x 0.5) metres wide and 5m + (2 x 0.5) metres long.

In other words, you would need to buy a liner that is 4m wide and 6m long.

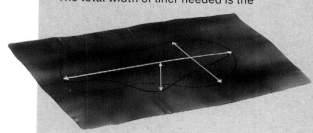

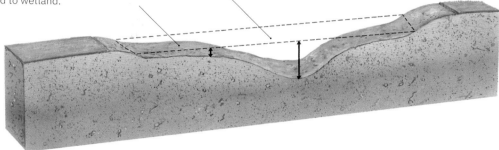

The edges of the pond slope gently, making a seamless transition from pond to wetland.

The majority of the pond should be less than 50cm deep to provide plenty of shallow water for marginal plants and the invertebrates that live there.

3 FIND A PLACE FOR YOUR TOPSOIL

Pile the topmost layer of soil some distance away from your pond. This soil layer is too nutrient-rich for your pond so make sure you keep it out of the way (mixed with compost, you could use it somewhere else in the garden). When you get to the less nutrient-rich soil underneath the topsoil, put this in another pile. This subsoil is more important for your needs. It will provide a less nutrient-rich substrate upon which you'll plant your pond plants.

4 REMEMBER YOUR PROFILE

In the final stages of digging, remember your pond's profile. You will want to keep a very shallow slope on at least one side of the pond, giving it a gentle transition from water to wetland. A central slightly deeper area (up to 50cm) may require a steeper slope towards the middle.

5 CHECK YOUR LEVELS

Using some boarding and a spirit level, check that each edge of your pond is level. Check and check again, adding or removing soil to achieve this. Getting your levels correct at this stage is really important because a slanted pond may expose the liner, making it more likely to puncture or be exposed to damaging UV radiation in time. Plus, aesthetically, it just looks bad.

6 PREPARE THE BASE Carefully go over your freshly dug pond base, removing any sticks, twigs or obvious stones. Once clear, add a layer of soft sand or old carpet. This will protect underneath the pond liner, helping ensure that it lasts as long as possible. Carefully unwrap your liner in the centre of the pond, ensuring that it is placed evenly and in the middle of where the pond will go. At this point, you can add some subsoil onto the liner as a substrate on which pond plants can colonise. This should form a layer no more than 2–3cm deep across most of the pond liner surface. If you are considering an area that will be boggy for much of the year, consider adding slightly more soil.

Once you're all done with the subsoil you can pile the rest near the most northerly side of the pond and cover with topsoil if necessary (this can be a sunny wildflower bank in due course). Now take a look at your prospective pond. It should simply be a liner covered in subsoil resting on a freshly dug hole. Don't be tempted to cut your liner down just yet. Let it rest there. You're now ready for some water.

7 JUST ADD WATER Ideally, your pond will need rainwater. Perhaps your best source of rainwater is to empty out water butts or to call on others for the contents of their own nearby water butts, decanting into buckets as necessary or attaching a hose directly to the water-butt nozzle. Though it may be a lot of effort traipsing back and forth with buckets, the rewards by way of diversity (and algae control) further down the line are more than worth it. If you have to use tap water and a hose, leave the tap water for a few days before planting to allow chlorine and other chemicals to dissipate. As the pond fills up, the water is likely to look very messy and cloudy. Fear not; over the coming week the silt will settle and the pond will start to become much clearer.

Once filled up you should have a good idea of how much liner you have left and how much you can cut back. At this point, you can cut it back to 60cm from the water's edge. You can cut it further later on if needed.

8 CHECK YOUR EDGES With the pond filled up, you will now have a good idea of how level your edges are. If there is a section from which water is trickling out, this is your moment to repair it by tucking soil or sand underneath the liner's edge to level things up. Carefully dig a channel around the perimeter and cover the liner with stones before covering it up with the subsoil turf you used earlier.

Planting up containers

Another option for planting up your pond is to use plastic baskets lined with hessian sacks, which can be filled with aquatic compost and planted up. These baskets are helpful if you want to buy ready-grown plants, and they can be moved around the pond if needed without making too much mess. If you plan on using baskets just remember to add some horizontal parts to your pond to stop them from tipping over.

Pre-formed ponds

Preformed plastic ponds are a hard-wearing alternative to flexible plastic liners, but they can be fiddly to install. A key thing to remember is to keep your spirit level handy at all times. Ensure a sturdy base using compacted sand so that the liner won't begin to slip downwards upon being filled.

9 **PLANT UP YOUR POND** The most wildlife-friendly ponds are those that are colonised naturally by plants in the local vicinity. However, many pond owners choose to add native plants directly to the soil themselves. Plants are discussed in more detail in Chapter 4, but a key thing to remember is that you want to start planting up your pond from the middle outwards so that when planting, you don't trample on any plants that you've already introduced.

Once the pond is lined you can add the plants. The shape of the pond will create areas for different types of plants.

10 **ADD WILDLIFE-FRIENDLY FEATURES** To help your pond reach its full wildlife potential you can add a range of companion features. These include log piles, rockeries, bird boxes, solitary bee nest-boxes, dragonfly perches and amphibian hibernacula. For more information on these features, see pages 55–57.

DIGGING A LARGER GARDEN WILDLIFE POND

What follows on these pages is a nine-step guide to creating a series of small clay-lined ponds that join up in winter to create a much larger pond. This is not a quick project. To build such a pond yourself will require you to have an understanding of the local hydrology of your site, and you may even need a number of permissions and licences depending on your whereabouts. Your statutory nature conservation agency (see page 155) will be able to help you with the planning of ponds of this type.

1 DON'T BE AFRAID TO GET EXPERT HELP
Once you have your plan, unless you are a qualified hydraulic-excavator driver, it is likely you will need to call on a pond specialist to take on the digging part of your project or to manage the project entirely. When seeking out possible contractors you will notice a range of prices. Ideally, try to find an expert with a good understanding of hydrology, and who has experience of making other large ponds for wildlife. Your local Wildlife Trust is likely to be able to provide advice and support with this, and may even know a trustworthy contractor. Indeed, some Wildlife Trusts may be able to offer their own specialist contractors for this exact job.

Before agreeing to work together with any contractor you will need to have an agreed timescale and fixed price – you don't want to be held liable if the contractor makes a mistake further down the line or if there is an unexpected increase in budget.

Digging out a large pond will create a lot of spoil, and you'll need to consider what you'll do with the excess. Removing soil from a site is very expensive so ideally you should consider making habitat features like butterfly banks or amphibian hibernacula with the excavated soil. As with small ponds, remember that diverting water from a river or stream will require special consideration and possibly a licence from your national statutory conservation agency (see page 155). The same goes for if you plan to introduce fish to your pond.

2 CONSIDER AND RECONSIDER WHERE YOUR POND WILL GO
Just as with small ponds, the location of your large wildlife pond is one of the most important decisions you'll make. Ponds can be affected by shade, aspect and slope, so getting this right now will make life easier further down the line. Remember that while your pond is being created the digger (or diggers) on site will make things very messy. Likewise, you may need access to your site for other large vehicles, including for spoil removal if you go down this route. If you are considering a large pond in a field, remember that the field itself can be vital habitat for rare species, particularly wildflowers. Destroying such habitats is not advisable. In situations like this, it may be that another field is more suitable.

3 WATER, WATER, WATER
Filling up a very large pond with a hose is not really a

feasible option, so you will be dependent on either inflowing water or groundwater to fill the pond. Most people will require a qualified hydrologist to assess how much water will be available to you through such sources. Specialist pond contractors should be able to undertake a hydrological survey for you. Remember that water inflows will often bring with them excessive quantities of nutrients that may, in time, lead to problems with algae. Planting a reedbed near to where polluted run-off water enters the pond will help to mop up some of the incoming nutrients.

4 REMOVE THE TOPSOIL
Because the topsoil won't hold water, you will need to scrape it back towards the margin of the pond so that you can get to the layer of clay underneath. Refrain from putting this topsoil near any inflows into the pond. The high levels of nutrients present in the topsoil may leach into the water, causing problems with algae in future years.

5 CONSIDER YOUR PROFILE
Some specialist pond contractors may have been involved in civil engineering projects, many of which favour symmetrical geometric shapes. So you may need to refer back to your original plan frequently to ensure that your pond's profile remains sloping and as natural-looking as possible during building. Many large ponds will begin filling up with water before the digging is finished. It may be that specialist contractors need to build temporary storage ditches to divert water away from the digging site. These may require their own permissions. Seek advice from your national statutory nature conservation organisation.

6 MAKE THINGS WATERTIGHT
Once you are happy with your pond's profile, you'll need to puddle the clay lining. Mostly, puddling is achieved by running heavy vehicles forwards and backwards over the base to make it waterproof.

7 CHECK LEVELS AND SHAPE
With the clay puddled, you should be able to see clearly how well the pond fits your original plan. Most diggers have specialist equipment to check depths and many will use a theodolite (a tripod with sights that can measure precisely the dimensions and levels of the pond's edge). You'll need to see for yourself that these levels line up and that the pond will fill up as intended on all sides before you sign off the project. It is far less costly to rectify problems now than after the pond fills.

8 ADD WATER
You may be required to unblock the inflow or divert a previously created storage ditch back into the pond to fill it up. This process may take many days, and it is very likely that during this time the pond will develop algal blooms, some of which will leave the pond almost pea soup-like in consistency. These algal blooms may last many weeks but should pass as invertebrate communities (particularly water fleas) develop and multiply to take on this numerous food source. It may take a month or two for your pond to settle.

9 PLANT UP YOUR BIG POND
In most large ponds, you could consider letting the pond colonise naturally. Over months and years, local seeds will arrive just as animals do, and your pond will begin to take its own shape that mirrors those of other ponds locally. However, some large ponds (for instance those in schools or urban locales) may be under pressure to have an early aesthetic 'wow factor' to them, so these ponds may need planting up artificially. A number of specialist pond plant providers exist online, many of which allow you to specify the size of your pond and will automatically provide a mix of native plants to suit. Some even provide ready-seeded coir rolls that can be attached to the edge of the pond. Remember that you will need to time your plant delivery right; if the plants arrive too early and your pond isn't ready, your planting efforts are likely to be in vain. You can learn more about which plants suit ponds in Chapter 4.

POND FEATURES
- NESTING SITES

Now it is time to consider the space around the pond. On the pages that follow, you can see ideas to help you encourage as wide a variety of animals as possible to your newly established pond, just by installing a range of accessory features.

COMPOST HEAPS
For all ponds, open compost heaps or compost bins can be a valuable accessory habitat in which animals like frogs and toads may hunt, and, in some cases, grass snakes may lay their eggs. Slow-worms (*Anguis fragilis*), too, are attracted to this feature, drawn in by the promise of good food and a place that stays warm even in cold weather. Many slow-worms may choose to birth their young in such a location.

SOLITARY BEE HOTELS
A number of bee species are attracted to the pond edge in their search for water or emerging pond flowers. For this reason, providing a nearby spot for them to nest makes good sense. In the wild, many solitary bees seek out holes made by wood-boring beetles for their nest sites. You can recreate these sites near your pond by adding a solitary bee 'hotel' – an easy-to-install habitat feature that consists of a box or other container filled with hollow plant stems, such as bamboo, which bees will seek out for egg-laying. Make sure your bee hotel has a slanted roof, so that it won't fill with water, and remember to keep it as dry as possible during winter. If all goes well, the developing grubs will emerge as breeding adults the following year. Drilling a range of small holes (between 2mm and 10mm) in nearby sun-drenched logs can also provide extra nesting sites for solitary bees.

purchased online. Ideally, bat boxes should be attached to a wall or a tree, between 3m and 5m above ground. Remember that bats need a clear flight path to enter and exit their boxes. Once installed, you can check for the presence of bats by looking for their droppings underneath the box. If you suspect bats are present, you will need to inform your country's statutory nature conservation agency (see page 155), which will provide you with the necessary advice and, if required, a licence.

Bee-nesters (above) can help bring buzz to your pond. Just remember to clear them out once a year. A DIY bird-box (right) or bat-box (bottom right) can be just as effective as those purchased from a shop.

BIRD BOXES In new dwellings, many walls lack ivy or other dense vegetation in which birds naturally nest. In these gardens, nest boxes can be a lifeline for garden birds while adding extra characters to the daily soap opera of your newly developing pond habitat. Many garden centres now stock bird boxes that encourage a range of different bird species. Armed with a saw, a hammer and some nails, you can also make a bird box yourself; there are a range of designs online. Larger open entrances are very attractive to robins (*Erithacus rubecula*), spotted flycatchers (*Muscicapa striata*) and wagtails. Pied flycatchers (*Ficedula hypoleuca*) and great tits (*Parus major*), coal tits (*Periparus ater*) and blue tits (*Cyanistes caeruleus*) prefer nest boxes with a small entrance hole (25–28mm).

BAT BOXES Many bats use large ponds as feeding stations. At night, they fly across the surface, picking off invertebrates (including midges and mosquitoes) as they emerge from their pupae. Bat boxes placed near your pond may end up becoming a regular roosting site. A range of designs exists and many can be

POND FEATURES – WARMING UP

SUBMERGED SOUTH-FACING GRAZING SITES In the early weeks of their development, tadpoles often cluster in the warmest parts of the pond. This is partly because warmer temperatures encourage their metabolism but also because the parts of the pond that receive the most sun have the most algae growing upon them. Long, flat rocks near the surface warm up particularly quickly and might in spring be home to hundreds of hungry tadpoles, each rasping their tiny beaks against the fast-growing algae. You can use rocks to create a miniature dry-stone wall under the water, which provides lots of cracks that tadpoles can flee to if disturbed. When hungry blackbirds appear at the water's edge you'll see the tadpoles quickly disappear, almost like magic.

BASKING SITES It's not only snakes that bask on spring mornings to warm up their bodies. Frogs do too. On spring mornings, frogs like to perch upon south-facing banks to catch a few early rays. South-facing rocks or stones, flanked by dense vegetation, mean that your amphibians can bask while remaining hidden from predators. If your pond happens to be within roaming distance of grass snakes, you may see these elusive and eloquent predators using the same basking spot later in the year to warm up after a cool dip.

SOLAR-POWERED WATER FOUNTAINS Because warm water is less proficient at holding dissolved oxygen, hot summers can mean that deep ponds may suffer from spells of mild deoxygenation. To remedy this in small garden ponds you could consider a solar-powered fountain or pump to push bubbles through the water. Though the impact will be relatively small, such additions can look aesthetically pleasing. There is a range of products on the market and, if maintained, some will last many years.

ABOVE **1** Tadpoles grazing algae from a sun-drenched rock. **2** A common frog enjoying the sun. **3** An oxygenating solar fountain.

POND FEATURES – SAFE PLACES FOR POND WILDLIFE

Garden ponds can become regular stop-offs for a number of visiting predators, including cats, crows and foxes. What follows are tips to give your pond creatures a fighting chance against these common and widespread predators.

LOG PILES The microhabitats that can form within a simple pile of logs are staggering. There are the gaps that appear as the bark peels back; the damp undersides, where snails and woodlice flourish; the top tier, where wolf spiders display and where dragonflies may perch; and the lower tiers, where toads wait to ambush prey in the shadows. But log piles are also vital safe houses for young frogs eager to seek shelter from cats, dogs and foxes. Many frogs reuse such sites again and again, rather like a primitive panic room.

ROCKERIES Like log piles, rockeries offer much by way of hiding places, and amphibians seeking locations to see out the winter months may even use them. Larger rockeries can be built upon bricks or hardcore covered in a layer of topsoil and more decorative stones. These understorey locations are attractive to amphibians because they remain frost-free over the winter months.

ROOT MAZES Complex root systems, like those created by yellow flag iris, can become important refuges for pond animals during the drawdown months, because they stay damp and the gaps and holes provide a variety of hiding places from predators. In the summer months, especially during extended dry periods, you will notice that newly metamorphosed froglets and toadlets gather there in their hundreds, waiting for heavy rains that allow them to run the gauntlet away from the pond without risk of their tiny bodies drying up.

POND FEATURES
- FEEDING SPOTS

Some garden features can encourage various animals to remain near the pond after their larval development. This means you can see them throughout the summer months. Some of these creatures, including adult frogs and toads, can end up performing an important pest-removal service for your garden.

CRITTER CURTAIN

For me, this is your pond's single most accessible and awe-inspiring companion feature. By dotting squares of old carpet around the sunny banks of your pond, gardeners can encourage hundreds of different invertebrates, which can be observed at any time of the day by carefully peeling back the carpet for a brief look. This can be especially exciting for kids who are eager to know what's about. The reason such sites can become so fruitful is because they are warmed by the sun, yet the carpet creates an impassable boundary to evaporation. The result is a warm, damp paradise that can be home to spiders, slugs, ground beetles, worms and even larger predators such as frogs, newts and slow-worms.

HEAT PADS

If you have any pieces of roofing felt lying around, you could consider placing them as specially cut tiles (50cm x 50cm) on the sunny side of the pond. This material warms up quickly, providing a free source of energy to reptiles like grass snakes that are out early in the season. Later in the year, such sites become home to burgeoning ant nests, whose flying life stages are an important food source for birds and bats in the summer.

Roofing material warms up quickly, providing a free source of energy for reptiles.

MINI MEADOWS

The value of wildflowers to your patch cannot be understated. Wildflowers provide an important food source to pollinators including butterflies, moths and hundreds of different flies, like charismatic bee flies in spring and hoverflies in summer. The presence of these insects complements nicely your pond's food web, bringing in aerial predators like flycatchers, swallows (*Hirundo rustica*) and bats. Many garden centres stock wildflower seed mixes that grow with minimal investment. Within years, what was once a bare patch of earth can become a dazzling rainbow of sights, sounds and insect activity.

1 Fireweed (*Epilobium angustifolium*); 2 Blackberry; 3 White nettle flowers (*Urtica dioica*); 4 Ragwort (*Senecio jacobaea*) with Cinnabar Moth caterpillar.

WONDERFUL WEEDS Though popular culture abhors them, many of the plants we classify as weeds have incredible value to nature. Their very presence in natural ecosystems over thousands of years has meant many animals have adapted specifically to feed upon them. But weeds can have other benefits, too. In some cases, their stinging leaves or thorny branches can be used as a makeshift barrier that stops investigative children or pets intruding on a part of the pond you might prefer to keep private. Here are some classic weeds you could consider encouraging near your pond.

• **Nettle** (*Urtica dioica*): Numerous butterflies and moths make use of this plentiful food source during the summer, including during their larval stage.

• **Dandelion** (*Taraxacum officinale*): Many moths, including the white ermine (*Spilosoma lubricipeda*) and the large yellow underwing (*Noctua pronuba*), use this colourful food plant. Plus, goldfinches (*Carduelis carduelis*) eat the seeds.

• **Rosebay Willowherb** (*Chamaenerion angustifolium*): As well as providing long-lasting colour to your pond edges, this nectar-rich plant is the food plant for caterpillars of the marvellous elephant hawk-moth (*Deilephila elpenor*).

• **Ragwort** (*Senecio jacobaea*): Incredibly, 30 insect species depend on this vibrant flowering plant. These include the stripy orange-and-black caterpillars of the cinnabar moth (*Tyria jacobaeae*). However, remember that ragwort is poisonous if eaten by animals, especially horses.

• **Bramble** (*Rubus fruticosus*): In terms of its value, bramble is the Swiss Army knife of weeds. Many moth caterpillars specialise on their leaves, and bees and butterflies feast on the flowers, while their berries become an important food source later in the year for birds and small mammals. The dense, thorny thickets that brambles produce as they grow are important refuges for sleeping hedgehogs and roosting birds. If you have brambles very near to your pond remember to cut them back so the thorns do not accidentally puncture the liner.

POND FEATURES FOR POND WATCHERS

POND-DIPPING PLATFORMS

There are distinct pros and cons worth considering when it comes to pond-dipping platforms. For starters, do you really need one? Pond-dipping platforms are useful if you want to reduce bank erosion, but overhanging platforms can give the illusion that the best animals are in deep water when, in reality, the richest parts of your ponds are the pond edges. Also, pond-dipping platforms can concentrate dipping in one area of the pond. If there are lots of people bunched together, clumsy trips and accidents with nets and dipping trays are far more likely.

On the other hand, there is no doubt that some pond owners like pond-dipping platforms because of the aesthetic charm they can provide. If you are considering installing a large pond-dipping platform, it's worth noting that this is not a job for the non-specialist. To add the foundations for the supports will require advanced planning, for instance, and a good understanding of the pond substrate – the last thing you want is a pond that leaks, making the pond surface unreachable with a net from the dipping platform.

ACCESSIBLE PLATFORM FOR POND DIPPING

Pond dippers with mobility issues may have trouble searching through busy pond-dipping trays that are on the floor. A raised shelf onto which pond trays can be placed (*see photo 2 opposite*) means that all pond users can inspect animals more closely.

DECKING AREA Adding a small area of decking near the pond can make it an attractive place to do the things you love, such as yoga, playing guitar, a bit of patchwork or, well, reading a book with a glass of wine. Under the decking itself, invertebrates and hunting amphibians may temporarily take shelter.

A POND HIDE A wildlife hide/shed (*see photo 1 below*) is an option for medium to larger wildlife ponds, particularly for pond watchers eager to capture impressive moments on camera. The wildlife hide will need to be on the side facing away from the sun, pointing at sloping sun-drenched banks and overhanging branches upon which dragonflies and damselflies may perch. Enthusiastic pond watchers sitting in wildlife hides have taken some impressive photos of bats hunting over water at night.

A HANDY GUIDE TO POND ANIMALS Many encounters you have with pond animals will happen when you don't have a wildlife guide handy. To remedy this, have a simple laminated guide to pond animals on a little piece of string near the pond. This will help you with trickier identifications that happen in the heat of the moment.

LIGHTING Solar-powered lights (*see photo 3 below*) add a decorative touch to ponds, encouraging us to spend time near them after dark. Indeed, they can act like a temporary moth trap, allowing pond watchers to encounter some of the emerging adult invertebrates coming out of the pond, including caddisflies, moths, mayflies and pollinating midges. Remember not to leave lights on for longer than is necessary. Light pollution may influence negatively the nocturnal goings-on in your pond.

CONNECTING UP YOUR PONDS

Wildlife corridors are the motorways upon which colonising animals will find your pond. By considering these corridors properly, you can improve the potential for wildlife in your wetland patch enormously.

WHAT ARE WILDLIFE CORRIDORS?

From above, animal habitats such as ponds are like tiles on a mosaic. For wetland animals to move between these tiles, they need opportunities for food and cover from predators. Your pond animals require a relatively clear run, without obstacles such as car parks or busy roads or pavements with high kerbs. Classic wildlife corridors include hedgerows and patches of woodland but they can include gardens, as long as they are not isolated by brick walls or impassable wooden fences with concrete bases.

Have a look at your garden viewed from above using a satellite mapping app or website. Highlight any ponds, lakes and/or streams within 500m. Now imagine life as a frog, toad or snake – could these creatures feasibly manage to get to your pond? If the answer is no, consider what you could do about that. Could you encourage some neighbours to make a 12–15cm access hole in the bottom of a fence panel? Could you lobby your local council for an amphibian tunnel to be built under a busy road? Could you talk to a local nature group about digging a pond in a nearby featureless recreational park? Could you encourage a local school to dig a pond to provide a 'stepping stone' between your pond and another, more distant wetland? All of these actions would help nature spread through neighbourhoods far more efficiently.

ABOVE/BELOW By using a simple hacksaw, some fenced gardens can be opened up to visiting hedgehogs and toads.

Ponds and climate change

As the climate continues to change across the UK, networks of ponds are likely to provide added value because they can support species as they adjust their distributions. But ponds also have another role here. Ponds may also become a weapon that helps us fight climate change itself. As ponds explode into life, day by day and hour by hour they pull carbon from the atmosphere as the organisms within them grow and reproduce. When these organisms die, the vast majority of the carbon contained within their bodies sinks to the bottom, forming a carbon-rich sludge that becomes locked away from the atmosphere. Recent research suggests that ponds may be capable of removing from the atmosphere a

mean average of 142g of carbon per square metre each year. By way of comparison, a square metre of woodland or grassland traps 2–5g of organic carbon each year.

ABOVE/BELOW New ponds in urban areas could one day help lock carbon away from out of the atmosphere.

WILDLIFE TUNNELS

In some cases, wildlife tunnels that run underneath roads can offer pond animals a valuable way of moving between networks of ponds, helping maintain the genetic diversity of the animals that live there and providing a well-connected habitat more resilient to local extinction. Most tunnel systems consist of a hard-wearing concrete tunnel, with fencing on either side to guide animals away from the road and towards the tunnels. However, there are downsides to tunnels – the vast majority of wildlife tunnels are costly to install and require yearly maintenance, for instance. Plus, international studies into their effectiveness show considerable variation in usage rates between sites, locations and wildlife species. However, at some sites and for some species – most notably threatened amphibians – wildlife tunnels between ponds are undoubtedly a

valuable option worthy of consideration in larger urban pond projects. The presence of these tunnels may even make or break an urban wetland site in terms of the wildlife it supports.

A CASE STUDY

Over four years, the wildlife charity Froglife monitored the amphibians that used a series of wildlife tunnels in Peterborough that linked a population of amphibians whose habitat had been bisected by a new road. To record amphibian movements, scientists pored over hours of video footage from camera traps put at either end of the tunnels to ascertain their usage, day and night. In 2019, the *European Journal of Wildlife Research* published the results.

In their case-study, road-fencing and tunnels reduced the likelihood of amphibians dying on roads as they moved between ponds and multiple new ponds were colonised by all four amphibian species present through the use of the wildlife tunnels. Most notable was how

newts used the tunnels significantly more in the autumn months as individuals moved between breeding sites, something not observed in frogs and toads. However, this is not to say that all amphibians completed a journey through the wildlife tunnels. According to the researchers involved, there were many occasions where amphibians performed apparent 'U-turns' while halfway through the tunnel in an apparent act of rejection. Nevertheless, overall, the effectiveness of the wildlife tunnel system on amphibians was positive. In fact, during the observation period, the scientists observed a rapid increase in the local great crested newt population, which they attributed in part to the new tunnel system.

BELOW/RIGHT Common toads are among several freshwater animals known to use wildlife tunnels.

Living landscapes and you

In recent years, all local Wildlife Trusts have been working hard to promote the value of connecting essential wildlife habitats through our Living Landscapes approach. Through Living Landscapes, we are attempting to join landscapes and environments to create a dynamic, complex and linked system which buffers local wildlife from widespread declines, including those caused by an increasingly temperamental climate. You can find out what's happening near you and how you can get involved by visiting your local Wildlife Trust.

SAFE POND DESIGNS

While ponds are a source of danger to unattended children (particularly those between one and two years old), you can take a number of steps to make them safer. Some of these steps are outlined here. There is further information about pond safety, particularly referring to keeping groups of pond dippers safe, on page 136.

CONSIDER YOUR DEPTH AND YOUR LINING
Though children can drown in very shallow water, the risk is greatly lessened in ponds without deep sections and those that have shallow, sloping sides rather than vertical edges.

ADD A FENCE AROUND THE PERIMETER OF YOUR POND
Many organisations, particularly schools, add a lockable gate and wooden fencing around the pond to ensure that groups can investigate the pond only in the presence of a responsible adult. If you choose this option, remember to keep

the key to your pond safe, and keep a spare in case the first key is lost. This will help ensure your pond doesn't become forgotten about and overgrown should a key be misplaced. Also, remember that you'll need to treat any wood used for the fencing.

POND GUARDS
Some companies offer specially fitted metal guards and cages over ponds to keep children out. On the whole, these guards come in two styles. One is a metal

mesh that is installed over the surface of the pond which, in some cases, can withstand the weight of an adult or child walking upon it. The other is more of a three-dimensional cage that is installed over the pond edge. Both styles can be made to measure and have options for little 'windows', which can be unlocked for getting a pond-dipping net in and out of the pond or for maintenance tasks. Though costly to install, pond covers like these can offer peace of mind and they are hard to vandalise. There are critics of pond guards like these, however. These critics argue that 'cages' send out a negative message about ponds and that by caging up ponds, we might inadvertently give the impression that all water is dangerous. It isn't, as long as you understand and eliminate the risks.

CONSIDER HOW TO MINIMISE RISKS RIGHT NOW

Though the idea of creating a formal risk assessment for the use of your pond seems a touch over the top, you might consider jotting down a quick table in a notepad, identifying risks in one column and outlining steps to remedy each risk in the next column. Risk assessments are discussed in more detail on page 137.

The RoSPA Water Safety Code

The Royal Society for the Prevention of Accidents (RoSPA) has a water safety code that is outlined here. Visit www.rospa.com and search for the Water Safety Code for further information, including information about signage.

SPOT THE DANGERS! Water may look safe but it can be dangerous. Learn to spot and keep away from dangers. You may swim well in a warm indoor pool but that does not necessarily mean that you will be able to swim in cold water.

TAKE SAFETY ADVICE! Special flags and notices may warn you of danger. Know what the signs mean, and do what they tell you.

GO TOGETHER! Children should always go with an adult, not by themselves. An adult can point out dangers or help if somebody gets into trouble.

LEARN HOW TO HELP! You may be able to help yourself and others if you know what to do in an emergency. If you see someone in difficulty, tell somebody, go to the nearest telephone, dial 999, ask for the police at inland water sites and the coastguard at the beach.

RIGHT Pond guards can be helpful around young children, but some critics argue that they give the wrong message about water safety.

MANAGING YOUR POND

AS NEW WAVES of algae-eating organisms arrive, your pond will mature slightly and become more and more predictable as the weeks pass. Within six months, if there is good light, most clean ponds should become crystal-clear. Almost like a child, your tempestuous pond is ready to grow.

Ponds really do age. As leaves and other plant materials build up and then die, a layer of detritus begins to build up on the pond bottom. Year by year, season after season, many ponds become shallower and more prone to drying up in the summer months because of this build-up of dead and decaying organic matter. This is totally natural for ponds. In fact, periodic drying is often a very good thing for the creatures that will call your pond home, reducing the diversity of predators and providing new niches (such as those in the drawdown zone) that some invertebrates specially seek out.

As a pond owner, you have a choice about how your pond ages. One option is that you accept this is the natural way with ponds, and you watch your pond and its inhabitants change over the years as it makes its inevitable journey towards becoming a bog garden. Or you take another option by trying to slow the ageing of your pond through some regular pond management.

In this chapter, you will discover a range of techniques for slowing your pond's decline, as well as some of the more drastic actions that pond owners can choose to return their pond back to a more youthful stage.

PROBLEM PLANTS AND ALGAE

Of all the pond's inhabitants, algae are perhaps the most maligned and misunderstood. Being the fastest and most energetic utiliser of the sun's light, as well as the most efficient organism at mopping up the pond's free nutrients, algae are perhaps the single most important resident in your pond for kick-starting food webs, particularly in the early years. Bluntly, without the algae in your pond, the water would have little by way of diversity. In this section we explore algae in a little more detail, outlining some suggestions for their management if required.

LEFT Under the microscope, blanketweed reveals itself to be hair-like chains made up of algal cells joined side-to-side.

RIGHT Each duck-weed is an island of photosynthesis capable of quickly reproducing when conditions allow.

WHAT ARE ALGAE?
The term 'algae' is rather a sweeping one. It includes literally thousands of largely unrelated species of single and multicellular organisms that occur in freshwater and saltwater across the planet. In the UK, there is estimated to be between 5,000 and 20,000 species.

Algae are united by one thing: each single or multicellular algal species is able to photosynthesise, provided it has access to varying small amounts of nutrients in the waters that surround it. Unlike plants,

multicellular algae usually consist of a single-cell type throughout the organism, meaning that most multicellular algae lack leaf-like structures. In ponds, the most advanced forms are the rather charming (and enduring) stoneworts, which resemble spindly freshwater seaweeds. In the UK, 33 species of stoneworts are known.

WHICH ARE THE PROBLEM ALGAE?

It is likely you have experienced ponds where algae have outcompeted everything else. These ponds often look green or blue (or sometimes red) and sometimes they can be made soup-like with the concentration of algal cells in the water. Other algae can have a hair-like consistency, which can quickly come to choke up ponds.

The problem algae can be broken down into two main types as follows.

- **Single-celled algae:** Blooms of free-floating algae often occur in new ponds or in the early days of spring. In dense numbers, these are the ones that can lead to water acquiring a soup-like consistency.
- **Filamentous green algae:** These very common green algae divide again and again, forming long filamentous strings, which wind throughout the pond like green slimy hair. The most common filamentous algae to most ponds are a clutch of species known to pond owners collectively as 'blanketweed' from the genus *Spirogyra*.

AND WHAT ABOUT DUCKWEEDS?

Many pond owners consider the tiny floating-leafed plants known as duckweeds to be an additional pain. Like algae, these tiny plants are quick to reproduce, meaning that a few hot days in summer is all it takes for them to take over the entire surface of a pond. The vast majority of duckweeds are native and widespread plants entirely adapted to a surface way of life. In fact, the larvae of some invertebrates depend on them, including that of the duckweed weevil (*Tanysphrus lemnae*), the small china-mark moth (*Cataclysta lemnata*) and the shore fly

In late summer, blanketweeds can reproduce with such efficiency that whole areas of the pond can become an apparent sludgy mess.

(*Ephydridae*). Duckweeds can be a great place for watching springtails, too, provided you have a magnifying glass. Springtails are otherwise referred to as Collembolans. At one point in prehistory, this quirky group of invertebrates rivalled the insects in their quest for world domination.

WHAT IS THE PROBLEM?
Though most duckweeds and blanketweeds are entirely natural phenomena, they can limit a pond when they grow with such voracity that they form a thick, deep curtain across the entire surface of the pond. This curtain stops light from penetrating into the deeper waters of the pond, meaning that submerged plants cannot photosynthesise. They can also limit the exchange of gases at the water surface, causing oxygen levels to plummet.

CONTROLLING PROBLEM PLANTS AND ALGAE

Algae and duckweeds prosper in environments where there are ample nutrients going spare. Being small and quick to reproduce, these organisms are simply faster at assimilating nutrients than many pond plants and animals. They are adapted to quickly flourish and once they are established they can be hard to nudge out.

Understanding your pond's free nutrients is very important in helping diagnose any problems you might be having with algae and duckweeds. Generally speaking, the more free nutrients your pond has drifting around, the more likely you are to have problems in the future. If you can remove, divert or limit the inflow of nutrients into your pond, you stand a better chance of keeping your pond in check.

In this section, you will find some suggestions for how you might go about doing this without the use of off-the-shelf tonics, some of which can have unpredictable results.

PULLING OUT BLANKETWEED

A tried-and-tested way to lower the amount of free nutrients in your pond is to use a rake to carefully pull out the blanketweed. Because of the winding and tangled nature of the blanketweed strands, one can pull in impressive handfuls of the stuff in a matter of minutes. Every handful you remove represents nutrients that can't be used by anything else in the water. In smaller ponds, you can also pull out blanketweed by dipping a stick into the water and turning it around gently, as if twirling spaghetti on to a fork. Once the blanketweed is out of the water, run it through a water-filled pond tray or a bucket before composting. This will give any animals you have inadvertently caught a chance to escape. This is especially helpful to tadpoles, which often get completely tangled up in blanketweed strands.

ABOVE **1** Blanketweed being pulled out of a pond spaghetti-style. **2** Duckweed being netted out. **3** Day-to-day management can help you keep on top of your jobs.

NETTING OUT DUCKWEEDS

Depending on the size of your pond, you may be able to carefully net across the surface water to catch and pull out rogue duckweed. This will keep your pond open for longer throughout the summer months. It is likely that during long sunny spells, the pond surface will cover up again with duckweed very quickly, so persistence is key. Leaving a net next to the pond will remind you to have a quick scoop every time you walk past.

With duckweed, you can leave your netted spoils in a pile next to the pond for an hour or so or, alternatively, run them through a water-filled bucket or tray. This will allow any pond animals (including tadpoles) an opportunity to wriggle, swim or crawl their way back into the water.

MAKE YOUR LOCATION FIT Many ponds with algae or duckweed problems are situated directly underneath trees or shrubs from which years of leaf litter has fallen. Each leaf that falls in, once broken down, is filled with nutrients that these problematic organisms utilise quickly for their speedy reproduction. To reduce the amount of leaves falling into your pond you could consider having nearby trees and shrubs cut back in the autumn.

STOP USING TAP WATER Tap water has within it nutrients that duckweeds and algae can quickly utilise for growth. Try and switch to using rainwater instead by hooking up a gutter pipe to a water butt or two.

ACCEPT AND COME TO LOVE AN OLD POND The simple truth is that blanketweed and duckweed are natural pond residents that colonise and thrive in ponds in the later stages of their existence when nutrient levels naturally become very high. Recurring problems with these organisms may mean that your pond is simply getting older and is on its way to becoming a bog garden, which is itself a very important habitat for invertebrates. Ask yourself whether you have space to add a new pond nearby to replace this one.

Barley straw

Inflows of nutrient-rich waters can cause some wildlife ponds to have more problems with algae (particularly blanketweed) than others. In cases like these, some pond owners seek out a natural remedy, such as a small bail of barley straw or a derived extract of barley straw that they add to their pond. Scientists are still investigating the mechanisms involved but it seems that as the barley straw is broken down in the pond, naturally occurring fungi release chemicals toxic to many forms of algae, essentially killing them off. Rotting tree bark (which releases special tannins as it breaks down) is known to be another option that can be used to control algae.

Barley straw can be a quick fix, but remember that there may be underlying reasons why your pond has problems with blanketweeds.

REPAIRING LEAKS

Punctures occur in ponds for a variety of reasons. Some prime culprits include fallen sticks, the paws of foxes or dogs, or years of sun damage that can lead pond liners to become brittle and less resilient with time. In the short term, you can avoid pond leaks by carefully managing your pond, covering up exposed parts of the liner with plants, or removing sticks or other sharp objects that may cause problems.

If you think you have a leak in your pond, it's worth double-checking to make sure. Ponds naturally become drier in the summer months, and sometimes what looks like a leak is simply the effect of warmer periods during which water evaporates at a higher rate. To assess whether it really is a leak, refill your pond at the end of the day and check the water levels the next day. Water levels can drop rapidly in hot weather due to evaporation and plant transpiration, but if it drops back to where it was within a matter of hours, the news is very likely to be bad: you have a leak. Next, you must locate it.

First, refill the pond and let the water level drop back down. When the pond water levels are at their lowest, work your way around the edge of the pond, carefully scanning an area approximately 10cm above and 5cm below the waterline. If no puncture is obviously visible while doing this, add a single centimetre of water to the pond and then fill a squeezy bottle with water and bright red food colouring. Squirt this in tiny amounts every few centimetres around the edge of the pond, taking a few moments to see whether the clouds of food colouring are drawn towards a nearby leak. If you find a leak using this technique, remember to keep going, because there may be more than one. This is time-consuming work but all of this effort will save you time in the long run. Once you are sure you've found the leak, drain the pond by a few centimetres and let the tear in the liner dry out fully before you begin your repair work. If needed, temporarily transfer pond plants (and animals) to buckets or a tank (see page 144) while the work is undertaken.

Powdered bentonite is one effective remedy for clogging up leaks in plastic-lined ponds. Pack the bentonite into the hole and gently refill the pond. The bentonite will expand upon coming into contact with water, sealing the gap. Aquatic sealants can be quite effective too, but you will need to sand down the damaged area carefully before you begin. Small holes can be repaired with a single blob of glue. Larger holes will require a 'sticky plaster' approach: use a piece of plastic liner that you've cut to size manually and glue this carefully over the area of the tear. These powerful glues and sealants can give off noxious fumes, so ensure you wear a face mask at all times if you choose to fix your leaky pond this way. You will need gloves for this work.

If you have an older pond that has lasted perhaps more than a couple of decades, your leak could be an early sign that the pond liner is beginning to age and is therefore becoming more brittle. If this is the case there may be more leaks in future, so you might wish to consider relining the pond rather than fixing recurring leaks again and again.

If you wish to add a new liner you should temporarily store animals and plants in buckets and drain the pond (see page 144). Remove the old liner and double-check the underlay is in good condition before adding your new liner.

Thankfully, leaks in clay ponds are easy to remedy. Mostly these occur because exposed clay has cracked or because the layer of clay has become too thin in a part (or parts) of the pond. You can use bentonite powder to repair clay-lined ponds. It is very effective, except in those ponds where the natural pH is too low or there are large concentrations of mineral salts. Many people choose to apply bentonite across the entire rim of the pond to reduce the likelihood of any further leaks in future years.

DROUGHT YEARS

Throughout the lifespan of your pond there are likely to be years where long spells of hot weather and sparse periods of rain mean that waters recede and your pond may even dry up. Though this may seem counter to your interests, these dry spells can be incredibly important for the pond wildlife. This is because many animals and plants are specifically adapted to a drying pond landscape. They can even flourish in these exact conditions. In fact, according to the Freshwater Habitats Trust, perhaps as many as half of all freshwater plants and animals can tolerate drought in some form. These include marginal species as well as aquatic plants such as water lilies, pondweeds and stoneworts.

BELOW A pond like this can survive prolonged periods of drought, provided there is an adequate drawdown zone.

Pond animals have numerous adaptations for dealing with dry spells. Some, like pond snails, retreat into their shells. Others, like water beetles, simply fly away to other ponds. Animals like freshwater shrimps and caddisfly larvae temporarily dig into the wet mud. There is even increasing evidence that frog tadpoles (including those of the common frog) can alter their development towards a fast-growing way of life to leave a drying pond more quickly. Smooth and palmate newts (*Lissotriton helveticus*) seem to be particularly drawn to seasonal ponds, as do great crested newts. This is probably because ponds that dry up seasonally are devoid of predators like fish that would normally hunt and prey upon their larvae.

WHEN TO STEP IN
• Though it is entirely natural for ponds to dry out occasionally, the sight of hundreds of struggling tadpoles (or fish, if you have them) in a tiny puddle at the bottom of your dried-up

pond should encourage you to take some form of action. Though it may be tempting to rehome these tadpoles somewhere else, it is much better to leave them where they are by adding a little more water. This will let them continue their metamorphosis in situ.

• If your pond is plastic-lined and has only recently been installed, you may have areas of the liner where sediment has not settled, and which lie bare and exposed to the sun as the pond dries up. This sun damage will age your liner. It may also be that animals, including badgers and foxes, will walk upon the liner in their nighttime investigations, increasing the likelihood of punctures.

• Lastly, if your pond is frequently used as an educational resource (e.g. a school pond used for pond dipping) you will of course require some actual water for this purpose!

As always, when topping up your pond, avoid nutrient-rich water from the tap, which may lead to the growth of problem plants further down the line. Instead run a hose from a water butt or use buckets, being careful to allow water to stand if necessary, so that it better matches the temperature of the water in the pond.

PONDS IN WINTER

Just as with drought, pond animals in temperate regions are surprisingly well adapted to deal with spells where an icy barrier covers the surface of the pond, sometimes for days or weeks.

The ways in which pond animals manage this vary. Some, like the nymphs of caddisflies, dragonflies, damselflies and mayflies, are able to create anti-freeze compounds in their cells, which stop their cell walls from rupturing upon being frozen. Others, like the water spider (*Argyroneta aquatica*), make a temporary bunker out of an air-filled empty snail shell, which they sleep safely within. Other air-breathers, such as backswimmers and water beetles, crowd around pockets of air trapped underneath leaves near the surface.

WINTERKILL Some animals do not always weather these cold snaps so well. The common frog is one such creature. Each year, long frozen spells give rise to a phenomenon in some ponds whereby the retreating ice brings the sight of one or two, or sometimes dozens in larger ponds, suffocated frogs. Many pond owners know this as 'winterkill'. Not all ponds experience winterkill but it can be a distressing sight for pond owners who, like me, love these charismatic garden residents.

The causes of winterkill are still being investigated but it is very likely something to do with the way in which frogs breathe. Frogs don't usually need to surface during the winter months. Instead, their outer surface becomes like an enormous lung, with oxygen passing directly through the porous skin and into the body and carbon dioxide diffusing outwards. However, during icy spells the water in the pond is blocked off from the atmosphere, reducing the capacity for gases to diffuse in and out of the water. In this sealed system where oxygen levels drop and carbon dioxide levels rise, some frogs fail to hold out for fresh oxygen and they sadly perish. How much of an impact winterkill has on

frog populations each year is unclear but the phenomenon is likely to have affected and impacted this species for hundreds of thousands of years.

The Freshwater Habitats Trust is currently researching winterkill with an aim to understand which ponds are most susceptible. Their advice to help reduce incidence of winterkill is to gently and safely sweep the surface of a pond clear of snow during cold spells. This allows light to penetrate through the ice, ensuring that pond plants can continue photosynthesising in the water. You will need to stand on the water's edge to do this, of course. For safety reasons, this action may suit smaller ponds only.

It may be that ponds deeper than they are wide are more susceptible to winterkill because dissolved oxygen mixes slowly in water and may fail to adequately mix across the lower regions of the pond where the frogs lie dormant.

OVERWINTERING TADPOLES

Just as some frog tadpoles can speed up their development to leave a drying pond more quickly, many tadpoles can slow their development to see out the winter months in a pond in which there is not enough food. This sees them metamorphose into froglets much earlier in the season the following year. In effect, these early froglets trailblaze the land ahead of the competition.

How often this happens in ponds across the UK is unclear. A 2004 study of garden ponds in Glasgow showed that overwintering tadpoles occurred in one in five ponds surveyed. The mechanisms that determine how this happens are of interest to scientists. It may be that as early as July, some tadpoles may slow their growth and divert energy towards survival into the next year.

ABOVE The ghostly figure of a frog that has suffered winterkill under the ice.

DE-SILTING PONDS

Naturally, your pond is likely to fill up with decomposing vegetation (and the bodies of tiny dead animals) as the years pass. Over time, this will form a layer of nutrient-rich sediment on the bottom of the pond. In many cases, this build-up is entirely natural and very much part of the life cycle of your pond. However, you may want to restore your pond to its former glory if you have specific species you are aiming to conserve, like great crested newts, or if the pond is used regularly for educational purposes for instance. This is incredibly mucky and time-consuming work, but the results the following year can be more than worth it.

DE-SILTING SMALL PONDS

For smaller ponds, you can de-silt by hand.

1 First, you will need to carefully sweep the water with a hand net and store ponds animals in temporary buckets or a glass tank filled with pond water (see page 144). You can store pond plants temporarily in plastic bags if necessary.

2 Once this is done, you can begin emptying the pond bucket by bucket. This is laborious work, so set aside lots of time or recruit some helpers. You could even make a chain of volunteers, passing buckets of muddy water back and forth. Bear in mind that emptying a pond may take more than 500 buckets' worth of effort so you may need regular breaks and to stay hydrated. A syphon tube is another option to get rid of the surface water.

3 Once you are down to the silt, remove the topmost layer into buckets – this layer of silt (surface silt) contains invertebrate eggs and plant seeds, so you'll need it when you're refilling the pond.

4 You are now ready to remove the rest of silt, which will likely be liquid in form and rather smelly. This excess silt is very rich in nutrients, so you can carefully apply it to parts of your garden if needed as a fertiliser. Be very careful not to pour the silt near the pond though – the last thing you want is for those nutrients to wash back into the pond with the next rainfall. Also, watch out for (and carefully remove) litter that has found its way into the sediment.

5 You should now have an empty pond. First, add your buckets of surface silt and then begin

refilling with clean water, ideally from water butts. Once the water is settled (which may take a few days if you are using tap water), you are ready to reintroduce your plants and animals.

DE-SILTING LARGER PONDS

De-silting larger ponds may be a task that requires specialist advice and support. Depending on access to the pond and the consistency of the silt needing removal, there are two popular options.

Excavators If the silt is solid or very boggy in consistency, a hired-in excavator can make quick work of de-silting a pond. Large excavators can reach 10m from the bank, but some more specialist diggers have a reach even longer (up to 18m). A separate dumper truck will be necessary for most ponds, to ensure that silt is deposited some way away from the pond so that it cannot wash back into the water.

Sludge pumps A sludge pump (*see photo above*) works like a giant vacuum cleaner that sucks up mud from the bottom of the pond. They come in a range of sizes and fit a range of budgets, and they are the attractive option for many pond owners. However, the sucked-up sludge is liquid in form and so produces much more by way of spoil, which cannot be pumped into streams, rivers or ditches. Often specially prepared storage lagoons are required when using sludge pumps. Again, specialist advice is recommended if going down this route.

When is the best time to de-silt a pond?

De-silting a pond is incredibly disruptive so you should choose a time of year when you consider that fewest animals will be affected and when the water is at its lowest level. Many experts argue that autumn is the best time to de-silt your pond. At this time of year, most pond invertebrates will be in their adult life stage or lying dormant as eggs ready for the following year. In addition, by October, most frog tadpoles will have left ponds as adults and there should not yet be any adult frogs returning to the water for their winter slumber.

REMEMBER THE RULES!

If there are protected species, such as great crested newts, in your pond and you intend to clear it out, you will need consent from your country's statutory nature conservation organisation. See page 155 for contact details.

YOUR POND
AND POLLUTION

In lowland Britain, a staggering four out of five ponds show evidence of having been negatively affected by pollution or intensive land-management practices. This makes the effects of pollution one of the biggest challenges to freshwater conservation at the current time. Though pollution may not be something that all pond owners will experience – particularly if your pond is supplied and topped up with clean rainwater – it is worth knowing the signs of pollution so that you can quickly take action should problems arise at any time.

WHAT CAUSES POLLUTION?

Three of the most significant freshwater pollutants involve nutrients (including phosphorus and nitrogen), pesticides (including garden pesticides) and heavy metals, such as lead and copper. These pollutants can wash into ponds from the surrounding catchment and may be particularly noticeable in those ponds neighbouring farmland or busy roads.

These groups of pollutants can cause a range of problems for ponds.

• Nutrients on their own are a natural ingredient of freshwaters but in excess they can fuel the growth of problem plants (including duckweed) and algae (including blanketweed) that quickly come to crowd out other pond residents.

• Although most modern pesticides (and herbicides) break down quickly in water, they are are toxic to plants and animals, and can cause short-term impact upon the pond's inhabitants.

• Heavy metals can have a direct toxic effect on animals and plants but they can also cause longer-term chronic problems. Because they fail to break down, these metals can remain in the pond silt for many years.

BELOW By taking water pollution more seriously, freshwater ecosystems in the UK could be given a massive, life-enhancing boost.

Signs of water pollution

Many polluted ponds will have all three of the following symptoms:

- uncontrollable and seemingly unstoppable growth of blanketweed and duckweeds;
- water that is thick and cloudy with green planktonic algae;
- water that lacks submerged plants (including pondweeds and hornworts), which have trouble growing in polluted waters.

If you think that your pond may be suffering from pollution, talk to a local garden centre. You may be able to take a chemical sample of the water that will shed more light on the problem.

POLLUTION SOLUTIONS When it comes to pollution, prevention is always better, cheaper and easier than cure. To put it simply, you will need to do all you can to stop pollutants on their journey through the catchment towards your pond.

This may require you making a physical barrier, such as a bank or a ditch, or a buffer zone of undisturbed vegetation between your pond and the source of the pollution. Reedbeds can provide a natural filter for some pollutants, effectively pulling nitrogen out of the water and allowing it to harmlessly disperse as gas. However, reedbeds do not provide a remedy for all pollutants. Plus, as they establish themselves, they themselves may require management to stop them taking over your pond.

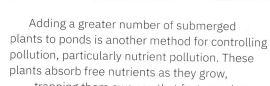

Adding a greater number of submerged plants to ponds is another method for controlling pollution, particularly nutrient pollution. These plants absorb free nutrients as they grow, trapping them away so that fast-growing algae and duckweeds can't use them.

If none of these methods work, then it may be worth considering de-silting the pond. However, de-silting is only worth doing if you can limit or completely stop pollutants from entering the pond in future. If you fail to tackle the source of the pollution, your efforts will always be in vain.

ABOVE Pond snails sound like a great way to remove excess nutrients yet, in truth, they serve only to move nutrients around the water body.

YEAR-ROUND MANAGEMENT JOBS

There are many jobs that can be undertaken at any time of year to ensure that your pond stays rich in wildlife for longer. It is worth bearing in mind that your pond plants and animals may not respond well to heavy disturbance. For this reason, consider management jobs sparingly and remember that a little bit here and there often goes a long way.

Removing litter Depending on where your pond is located, the sad truth is that litter is likely to blow in from time to time. In most cases, litter can be carefully netted out and recycled accordingly.

Cutting back brambles Maintaining dense areas of foliage around the pond will provide a useful hiding place for larger animals, including hedgehogs and amphibians. Watch out for brambles sprouting too near the pond. Their sharp thorns can puncture a plastic liner and so you will need to regularly cut them back.

Removing fallen leaves from the water The more leaves that fall into your pond, the more nutrients become available for fast-growing algae and duckweeds. Regular sweeps with a pond-dipping net in autumn to collect the fallen leaves can limit the amount of nutrients entering your pond, meaning that the pond is less likely to need cleaning out in future years. Carefully check your netted leaves, however, swilling them around in a bucket of clean water to ensure that no aquatic invertebrates (particularly water hog lice) are accidentally removed.

Cutting back surrounding plants in autumn

The plants around the edge of your pond are likely to need cutting back in the autumn months so that they do not die and fall into the water, where they may release problematic nutrients.

Netting out pondweeds

Throughout spring and summer many ponds experience dramatic growths of duckweeds (which can be carefully removed with a net) and blanketweeds (which can be removed with a small rake or long stick). As little as 60 seconds each day is all it takes to keep ponds from becoming too overrun. Just remember to rinse your removed problem plants through a bucket of pond water before composting them. This gives time for any trapped animals to escape, which you can then put safely back in the water.

Sweeping off snow

To allow pond plants to continue photosynthesising during cold spells, you may wish to carefully sweep snow from the icy surface of your pond. This will keep the pond oxygenated, ensuring that sleeping frogs in the pond, for instance, do not suffocate.

Varnishing pond platforms or signage

If your pond has a small decking area or platform, you will need to treat it each year to ensure that the wood does not become rotten. Do this on a dry day with no rain forecast for at least four or five days. This will reduce the likelihood of chemicals from the varnish entering the pond. You may wish to nail chicken wire to the surface to keep the platform from becoming too slippery.

Cutting back overhanging trees

As well as dropping leaves, trees can also shade out some ponds over time. Cutting back some branches in autumn can help. Though it may be tempting, don't go too far by removing trees entirely; the animals in an established pond may actually thrive there because of the presence of a large tree, rather than in spite of it. Suddenly changing things may disrupt a healthy and otherwise fully functioning wildlife pond.

PLANTING UP YOUR POND

A TRULY WILD POND should be left to colonise naturally, particularly if the pond is situated near existing freshwater habitats such as ponds, lakes or streams. In time, naturally occurring algae will blossom before the arrival of flowering plants blown in from seeds. Each of these plants will come and flourish in successive waves of activity. Within a few years, both submerged and emergent pond plants will have spread and prospered, with new plants appearing each year. Some of these natural ponds may even come to have more diversity by way of their plant communities than a pond planted up artificially. Pond animals and plants are supremely adapted for travel, after all. Colonisation is in their DNA.

Understandably, however, not everyone has the patience to let plants colonise naturally. Some wildlife ponds need to look nice straight away, for instance if they are in a public place like a school. Other pond owners may want a bit of colour as soon as possible, to brighten their garden and encourage visiting pollinators. For these reasons, many pond owners prefer to speed things up a bit by planting pond plants themselves.

Planting up ponds isn't quite like planting in a flowerbed. Being three-dimensional in their nature, there are different zones to consider in a pond, and pond plants differ in their needs. Some pond plants prefer deep water, for instance,

while others prefer the shallows. Some plants like the pond edges whereas others prefer the surface.

In this chapter we outline which native plants grow best in each pond zone, as well as considering in more detail the native pond plants that encourage the broadest spread of wildlife, particularly invertebrates.

SOURCING POND PLANTS

It's worth reiterating first of all that plants really do find their way naturally to ponds as seeds or tiny fragments of leaves. If you live within a few hundred metres of a freshwater habitat, it is likely that many plants will blow in on the air or be brought in through the feathers or droppings of birds. Just as with colonising animals, there is a certain pleasure in seeing which plant species arrive and how they take to your pond. Some plant species will colonise and thrive even within months of their arrival. Indeed, new ponds like

Increasingly, many garden centres offer a range of native pond plants, but be wary of poor labelling. Check and check again. Go for native plant species every time.

these can be vital for dragonflies such as the four-spotted chaser (*Libellula quadrimaculata*) and common darter (*Sympetrum striolatum*), which prefer bare ground or do not compete well alongside other species. The key message? Natural is best, every time.

Of course, for those not near freshwaters or those living in more urban locations, the wait for plants can be long and, well, a bit tedious. For these pond owners particularly, sourcing pond plants can become a naturally desirable second-best option. So where is it best to acquire good-quality native pond plants?

Clearly, garden or aquatic centres are one accessible option. Indeed, in recent years many of these shops are now stocking native species and are more understanding of the need for native planting. Also, all garden and aquatic centres are now banned from selling the worst and most invasive non-native pond plants, many of which cause millions of pounds' worth of damage each year due to their proliferation in the wild. But buying from some garden centres can be problematic. Garden-centre stock can sometimes be mislabelled or misnamed or, worse, can unwittingly be home to a number of invasive pond plants that could accidentally hitch a lift into your pond with your purchase. For many pond invasives, a single leaf is all it takes for transmission, so a single purchase can give you more than you bargained for.

A better and often more rewarding option for planting up ponds is therefore to take cuttings and seeds from natural ponds within 20km of your own. You will require permission from a land owner or another garden owner to do this, of course, but by keeping it local you can better guarantee that plants will survive and that you're helping to keep the gene pool of local wetland species intact and as natural-seeming as possible.

PROPAGATING POND PLANTS

The best time to divide pond plants to take some back to your pond is spring. This is when plants are starting to grow, ensuring the longest possible growing period in your new pond before the cold winter arrives. Propagating plants can be undertaken in a variety of ways, but in all cases, plants will need to be kept wet during transportation to the new pond. Here are some examples of the more common pond plant species that propagate well.

Marsh-marigolds Many marsh-marigolds (*Caltha palustris*) form dense clumps of roots that can be broken up by hand or with a fork before being brought up. Look for the healthy-looking outer edges of these clumps for planting. The inner areas may well be old and exhausted.

Water lilies Established water lilies (*Nymphaeaceae*) form dense mats that you can net towards the water's edge and cut into smaller chunks using a handsaw before you transport them to your pond. This is messy work, but often water lilies take well when divided in this way.

Creeping plants Plants that spread along the shallows, such as bogbean (*Menyanthes trifoliata*), water mint (*Mentha citrate*) and lesser celandine (*Ficaria verna*) can be easily pulled up with their short roots intact. These roots are easy to transfer and they also take to new ponds well.

Bulbs Plants like fritillary grow from bulbs, which can be carefully dug up and split exactly the same as if they were daffodil bulbs in autumn.

Bulbils On flowering rushes, there is a structure a little like a bulb, called a bulbil, at the base of the roots. Ensure each plant is removed with the bulbil intact and plant these in wet compost at home to let new, healthy roots develop. Once roots are firm, you are ready to introduce the flowering rushes to your pond edge.

Things to remember
• If you are sourcing pond plants, keep it native and keep it local.
• Carefully wash incoming plants before planting in case of accidental movement of invasive species.
• Avoid using nutrient-rich topsoil unless absolutely necessary. Most pond plants grow well on subsoils such as sand or clay.

PLANTING ZONES

For the purposes of this book, we have divided pond plants into three types and, therefore, three zones:

1 WETLAND AND POND-EDGE ZONE

This is often the most colourful area of a pond. It is a gathering place for young amphibians leaving the pond and provides a temporary shelter for invertebrates moving in and out of the water. In almost every way, this is your pond's red carpet.

2 SHALLOW-WATER ZONE

This zone can be a busy one for both tall and low-growing plants. The roots and complex tangle of stems and leaves provides important habitat for water beetles as well as predator-proof cover for tadpoles. This zone includes the 'drawdown zone' – an area capable of drying out (and that almost readily does so) for weeks or months in summer. This is another very rich area for pond diversity.

3 DEEP-WATER ZONE

For invertebrates, this pond zone is a bit like the open ocean. Lacking much by way of food and shelter, diversity is comparably low in this zone, although these areas can be important for fish and aquatic bugs. Deep-water planting can add structure here in the form of dense leaves, such as rigid hornwort (Ceratophyllum demersum) or long trailing stems like that of the water lily that, when numerous, offer an underwater aesthetic not unlike that of a bamboo forest.

Of the deep-water plants, there are two further subdivisions: those that are submerged and those that have floating leaves. It's worth noting here that the concept of specific 'plant zones' is a little misleading. In reality, there are crossovers between these zones. Yellow flag iris, for instance, can grow in standing water or the water margins, and many submerged water plants can grow in shallow waters as well as in deep water.

How to plant your pond plants

Though planting a pond can feel daunting to some, it really needn't be as difficult as many guides like to insist. In fact, you can plant most pond plants by hand directly into the substrate with no tools at all.

Here are some important points to remember before planting:

• The most critical moment in planting up ponds is transferring the plants to your pond without them drying up or being damaged. While in transit, keep plants bagged up or in buckets and keep them wet at all times.

• Avoid fertilisers and resist the temptation to add extra topsoil while planting. These will add unnecessary pollutants to the water, possibly encouraging the growth of problem plants and algae.
• Some pond plants can drown! Remember that with emergent and floating-leaf plants at least some part of the plant needs to be in contact with the air at all times.
• Planting won't always be successful. If something doesn't take, try something different or try planting the same plants in a different zone, perhaps nearer the pond edge.

Container planting

Many pond owners prefer to plant up their ponds using plastic containers lined with hessian sacks and filled with aquatic soil. The benefit of containers like these is that they can be moved in and out of the water more easily and replanted with little fuss, even (if necessary) seasonally. Baskets can also be a handy way to restrict the growth of plants with fast-spreading roots, such as water lilies and water mint.

The downside of container planters is that they can sometimes topple over if the pond floor isn't level, making a cloudy mess of the water. Note that the plastic casing can sometimes become brittle and tear after a few years. Sometimes roots can burst through old containers before spreading through your pond.

RIGHT You can prop up plastic planters using large pebbles or cobble stones to stop planters falling over and making a mess.

DEEP-WATER PLANTING – SUBMERGED PLANTS

In the UK, water milfoils, water starworts and pondweeds are some of the most common aquatic submerged plants. The good news is that these are easy plants to propagate from friends or neighbours with ponds. In the spring and summer, submerged pond plants can take quickly if your pond is suitable. In fact, some submerged pond plants can take so readily that they can quickly dominate your pond. If this happens to you, fear not: submerged plants respond well to being pulled out by hand or with a rake or stick.

Submerged pond plants bring two obvious advantages. First, they will help to oxygenate your pond. Second, every species of native submerged pond plant will provide important microhabitats within which invertebrates will come to breed, feed or seek shelter. The classic native species of submerged pond plants include spiked water-milfoil (*Myriophyllum spicatum*) or whorled water-milfoil (*Myriophyllum verticillatum*), both of which require non-alkaline waters. These are fast-growing species and each brings a certain delicate beauty to your pond. Water starworts are another option. There are multiple species of water starwort and all species specialise to varying degrees in still, slow-flowing or fast-running waters. For all of these submerged pond plants, you can add around three or more different species

to see which take best. It is highly likely that not every submerged plant you add to your pond will take. Many deep-water pond plants can be quite fussy about nutrients and the pH of the water, for instance.

You can easily plant most species of submerged pond plants simply by throwing cuttings into sunny regions of your pond. However, you may need to plant more sensitive species such as water milfoils into the substrate.

Opposite is a handy table of native submerged plant species that can take well in wildlife ponds. Remember: you may wish to plant your deep-water submerged plants first, so that you don't trample newly establishing plants around the edge of the pond while planting the deepest parts.

RIGHT **1** Curled pond-weed (*Potamogeton crispus*) **2** Soft hornwort (*Ceratophyllum submersum*) **3** Spiked water milfoil (*Myriophyllum spicatum*) **4** Fennel pondweed (*Potamogeton pectinatus*)

NATIVE SUBMERGED PLANT SPECIES FOR PONDS

species	depth to plant	where to plant	notes
Common water starwort (*Callitriche stagnalis*)	Up to 1m	Full sun or partial shade	Adapted to slow-flowing streams but also grows in still water
Curled pondweed (*Potamogeton crispus*)	Up to 75cm	Full sun or partial shade	Wide-ranging. Needs clear water
Rigid hornwort (*Ceratophyllum demersem*)	Up to 1m	Full sun or partial shade	An adaptable oxygenator. Can take quickly
Soft hornwort (*Ceratophyllum submersum*)	Up to 30cm	Full sun	Slightly less adaptable than rigid hornwort
Fennel pondweed (*Potamogeton pectinatus*)	Between 0.5cm and 2.5m	Full sun or partial shade to shade	Can tolerate more turbidity than other submerged plants
Spiked water-milfoil (*Myriophyllum spicatum*)	Up to 2m	Full sun or partial shade	Prefers clear, alkaline waters
Willowmoss (*Fontinalis antipyretica*)	Up to 2m	Full sun or partial shade	Can grow in fast-flowing water and still water
Common water-crowfoot (*Ranunculus aquatilis*)	Up to 50cm	Full sun	Likes still or slow-flowing water

Good plants for newts

In springtime, male and female newts of three species visit ponds, looking for places to lay their eggs. Unlike the common frog, which lays its eggs in vast blobs, and the common toad, which lays its eggs in long strings, female newts lay their eggs one by one. Using her hind legs, the female carefully wraps each of these individual eggs in a leaf or stem, where a tiny newt larva will grow alone, in safety. Newts find some pond plants especially effective for the purpose of egg-laying. These plants include:

• Grasses (*Glyceria* spp).
• Water mint (*Mentha aquatica*)
• Water forget-me-not (*Myosotis scorpiodes*)
• Overhanging nettles, brambles and even geraniums with leaves drooping into the water.

If your pond lacks these plants, don't worry too much. Many female newts will find something to lay their eggs on, including decaying leaves and grasses or even sticks and fallen branches.

DEEP-WATER PLANTING – FLOATING-LEAF AND FREE-FLOATING PLANTS

Many plants are rooted in the pond substrate and possess long trailing stems upon which leaves sprout that float on the surface of the water. The most celebrated of these are the water lilies. In the UK, our native water lily species flower white or yellow during the summer months, depending on the species. In the right circumstances, these floating-leaf plants can spread right across the surface of your pond, particularly in warm years. Thankfully, water lilies respond well to some considered cutting or pruning.

Historically, water lilies are species associated with ponds in floodplains or extensive wetlands, so they may not be a natural resident of all ponds. They are certainly beautiful, however, hence their popular appeal.

The best way to acquire water lilies is to think local; ask friends and neighbours first before you consider heading to your local garden centre. A simple option for acquiring water lilies is to collect seedpods in autumn, which you

BELOW **1** Broad-leaved pondweed (*Potamogeton natans*) **2** Frogbit (*Hydrocharis morsus-ranae*) **3** White water-lily (*Nymphaea alba*) **4** Water soldier (*Stratiotes aloides*)

NATIVE FLOATING-LEAF AND FREE-FLOATING PLANTS FOR PONDS

species	flowers	where to plant	notes
Broad-leaved pondweed (*Potamogeton natans*)	Small, green-coloured flowers between May and August	Full sun in water 0.3–3m in depth	Flower spikes protrude from the surface of the water. Has distinct underwater and above-water leaves
White water-lily (*Nymphaea alba*)	Striking white surface flowers between July and August	Full sun in water up to 3m in depth	Likes still water. Can grow profusely, so planting baskets recommended
Yellow water-lily (*Nuphar lutea*)	Large, yellow flowers between June and August	Full sun in water up to 1.5m in depth	Likes still water. Also capable of growing profusely
Water violet (*Hottonia palustris*)	White/bluish flowers from elegant stems that protrude from the water surface, May/June	Full sun	Easily confused with water milfoil when not flowering
Frogbit (*Hydrocharis morsus-ranae*)	White buttercup-like flowers between July and August	Full sun	Grows well in slightly alkaline waters. Looks a bit like a miniature water lily
Common water-crowfoot (*Ranunculus aquatilis*)	Capable of forming dense mats of white buttercups that flower throughout spring and early summer	Full sun in water up to 50cm in depth	Fairly adaptable to slow-flowing and still waters. Traditionally known from alkaline ponds
Water soldier (*Stratiotes aloides*)	White flower protruding from each rosette of leaves in July/August	Full sun in shallow or deep water	This plant is rare in the wild but common to garden ponds throughout the UK because of its adaptable nature

can plant in your own pond simply by throwing them in the water and allowing them to rest on the pond substrate. Just remember to keep the seeds damp while transporting them. Another option, particularly during early spring, is to cut off a section of the rhizome root mat of someone else's water lilies before transferring them to your own pond. Try to cut off the budding rhizome roots, to ensure that you stand the best chance of them taking.

Water lilies are not the only floating-leaf plants worthy of your attention. There are others, like water soldier (*Stratiotes aloides*), common water-crowfoot (*Ranunculus aquatilis*) and water violets (*Hottonia palustris*), all of which have leaves that provide resting spots on the surface for all manner of invertebrates, including dragonflies and damselflies. Their floating leaves also provide crucial shade from the sun for animals under the water, particularly fish and newts. Some floating-leaf plants are vital to the life cycles of invertebrates. The china-mark moth caterpillar, for instance, is a specialist of broad-leaved pondweed and even makes a case out of its leaves.

Our planting guide above gives examples of native floating-leaf and free-floating plants suitable for a vibrant wildlife pond.

SHALLOW-WATER ZONES

You should aim for your shallow zone to be as busy as possible. Like a coral reef, this region contains the most niches and therefore the most opportunities for invertebrates to exploit. In time, the shallow zone can become the beating heart of your pond's food webs. This is a highly competitive zone, however, with plants often jostling for position to occupy such prime real estate. Some careful consideration can help you maximise the species of pond plant that flourish here. Just remember that not everything will take first time.

WETLAND GRASSES The shallow-water zone can become an important part of the pond for wetland grasses, including floating sweet-grasses (*Glyceria* spp.) and creeping bent (*Agrostis stolonifera*). These grasses sweep along and underneath the pond surface, providing much-needed architecture within which animal life thrives. In fact, seasoned pond dippers often net near such plants because they know so many water beetles, pond bugs and newt larvae may be found there. Sweet grasses aren't commonly sold in garden centres. You may need to gather these yourself, should you have permission from the land owner.

REEDS AND BULRUSHES In the early years of your pond you may discover common reed and bulrush (*Typha latifolia*) naturally colonising your pond. Bulrush, particularly, is an effective coloniser of new ponds. Though these shallow-water plants do provide good aquatic habitats for many animals, they can quickly take over a wildlife pond. In addition, their ability to fix nitrogen can also be a problem because added nitrogen in a pond can lead to excess growth of duckweeds and algae further down the line. If you have a small pond, you may choose to limit the growth of bulrushes and reeds, cutting back the roots well beneath the water surface before pulling these roots and shoots clear from your pond. Though some wildlife pond guides encourage the planting of spearworts, with their beautiful yellow flowers, these too can be very prolific and not suitable for all garden wildlife ponds.

Many of the native shallow-water plants included in this list are commonly found in garden centres, but working from seeds or from cuttings is recommended as the cheapest and most satisfying option.

BELOW **1** Amphibious bistort (*Persicaria amphibia*) **2** Arrowhead (*Sagittaria sagittifolia*) **3** Bogbean (*Menyanthes trifoliata*) **4** Yellow flag iris (*Iris pseudocorus*)

NATIVE SHALLOW-WATER PLANTS FOR PONDS

species	plant height	flowers	where to plant	notes
Amphibious bistort (*Persicaria amphibia*)	30–70cm	Large pink flowers between July and September	Sunny banks or floating on sunny surface waters	This adaptable plant can grow in submerged waters as well as moist soils
Arrowhead (*Sagittaria sagittifolia*)	30–80cm	Small, white flowers between July and August	Shallow waters in full sun	Spectacular surface leaves provide good perches for aerial predatory insects
Bogbean (*Menyanthes trifoliata*)	10–30cm	Fluffy, white flowers between May and June	Full sun, can also be grown in deeper waters	Capable of prolific growth but easy to cut back
Brooklime (*Veronica beccabunga*)	20–30cm	Small blue flowers between May and September	Partial shade in shallow water no more than 30cm deep	Provides thick cover for invertebrates
Floating sweet-grass (*Glyceria fluitans*)	25–90cm	Silvery white, from May to August	Full sun	Trailing grass through the water provides excellent architecture for invertebrates
Flowering rush (*Butomus umbellatus*)	150cm	Pink flowers between July and September	Full sun	A firm favourite among wildlife gardeners
Water forget-me-not (*Myosotis scorpioides*)	15–30cm	Delicate blue flowers between May and September	Partial shade in shallow water no more than 30cm deep	The leaves of this plant create rafts on the water surface upon, which invertebrates and even froglets can gather
Water mint (*Mentha aquatica*)	15–60cm	Clusters of pink flowers from July to October	Sun or partial shade in shallow water no more than 30cm deep	One of more common plants upon which newts lay eggs. Capable of prolific growth
Water plantain (*Alisma plantago-aquatica*)	30–90cm	Delicate white spiky flowers from June to August	Sun or partial shade in water no more than 50cm deep	This adaptable plant is easy to plant straight into the soil substrate
Yellow flag iris (*Iris pseudacorus*)	45–150cm	Large protruding yellow flowers between May and August	Full sun, grows in water or on the pond bank	Attractive flowers frequented by a host of pollinators. Complex root mats can provide useful temporary housing for froglets and toadlets

WETLAND AND POND-EDGE ZONES

Your pond's wetland zone is another area capable of producing impressive colour and aesthetic charm. Take, for instance, the bright pinks of valerian (*Valeriana officinalis*), the blues of bugle (*Ajuga reptans*) and the gorgeous yellow flowers of lesser celandines (*Ranunculus ficaria*). All of these flowers provide a nectar-rich food source for pollinators including bees, flies, beetles, butterflies and moths.

Many of these plants can be grown easily from hand-collected seeds or, increasingly, through online gardening suppliers. Just remember to ask for permission from land owners if taking cuttings or seeds, and if venturing online, make sure you trust your sources.

Among the most colourful wetland plants are those that colonise ponds easily. Many wildlife ponds will have dotted across their edges numerous examples of marsh-marigold (often called kingcup), purple loosestrife (*Lythrum salicaria*), which flowers right into autumn, and great willowherb (*Epilobium hirsutum*). Each of these wetland plants can seed easily and spread widely. Even the most remote ponds are likely to eventually end up with some of these species, such is their prowess for spreading to new places.

In the years after planting, it is likely that a high density of plants will thrive around your pond's edges. Be aware that this dense vegetation may limit your access to the pond from all sides, so you might need to manage your pond edges a little every now and then. Carefully considered stepping stones or decorative slabs dotted around the pond edge can be one way to access the pond throughout the year, regardless of how dense the vegetation gets. This ensures you can undertake management tasks such as cutting back fast-growing plants, including brambles and sapling trees, should they occur. These stepping stones can also be handy places from which to pond dip throughout the year.

RIGHT **1** Bugle (*Ajuga reptans*) **2** Great willowherb (*Epilobium hirsutum*) **3** Purple loosestrife (*Lythrum salicaria*) **4** Common valerian (*Valeriana officinalis*)

WETLAND AND POND-EDGE PLANTS FOR PONDS

species	plant height	flowers	where to plant	notes
Bugle (*Ajuga reptans*)	10–15cm	Deep blue flowers between April and June	Partial shade	A colourful option for shadier ponds
Common valerian (*Valeriana officinalis*)	30–120cm	Rounded clusters of white/pink flowers in June/July	Full sun or partial to full shade	Also found in wet meadows
Creeping jenny (*Lysimachia nummularia*)	Less than 10cm	Yellow flowers between May and August	Full sun	This sprawling-plant provides impressive ground cover
Cuckoo flower (*Cardamine pratensis*)	30–60cm	Pale pink flowers in late spring and early summer	Full sun or partial shade	Flowers at the same time as the eponymous bird calls
Devil's-bit scabious (*Succisa pratensis*)	60–110cm	Lilac/purple flowers between July and October	Partial shade	Also provides autumn food for birds
Great willowherb (*Epilobium hirsutum*)	120cm	Protruding pink flowers in July and August	Full sun or partial shade	Establishes with ease and grows quickly
Lesser celandine (*Ranunculus ficaria*)	5–25cm	Yellow flowers from March to May	Partial to full shade	One of the earliest wetland plants
Marsh cinquefoil (*Potentilla palustris*)	10–45cm	Pink/purple flowers between June and July	Full sun or partial shade	Good option for more acid soils
Marsh-marigold (*Caltha palustris*)	30–40cm	Bright yellow flowers from March to May	Full sun to partial shade	Grows predictably well next to most ponds
Meadow buttercup (*Ranunculus acris*)	10–30cm	Yellow flowers between June and October	Full sun or partial shade	Comes back well from regular disturbance, including grazing
Meadowsweet (*Filipendula ulmaria*)	60–120cm	White flowers from June to September	Full sun or partial shade	Superb ground cover for amphibians and mammals
Purple loosestrife (*Lythrum salicaria*)	60–120cm	Spiky purple flowers between June and August	Full sun or partial shade	Well loved by wildlife gardeners for attracting pollinators
Ragged robin (*Lychnis flos-cuculi*)	30–50cm	Long and feathery pink flowers between May and August	Full sun or partial shade	Common to wet meadows as well as ponds
Yellow loosestrife (*Lysimachia vulgaris*)	60–150cm	Numerous yellow flowers between July and August	Full sun or partial shade	Sudden vast flowers can prove immensely attractive to pollinating insects

INVASIVE POND PLANTS

Animals won't be the only visitors to your pond. In time, it may be that invasive non-native pond plants turn up too. Invasive pond plants cause enormous problems for the UK's waterways. Their fast growth can block streams and rivers, which in severe cases can lead to flooding. They can find their ways into ponds as seeds, through transmission via animals or through accidental or deliberate transmission via garden centres or gardeners. In many cases, a single leaf moved into a new pond is all it takes for these problem plants to take hold and flourish.

Four highly invasive and troublesome non-native pond plants are detailed on this page. Watch out for these in your own pond or those nearby.

Australian swamp stonecrop/New Zealand pigmyweed (Crassula helmsii)

Once established, this non-native pond plant is capable of blanketing the surface of ponds with its leaves. This limits photosynthesis in the pond and causes knock-on impacts further up the food chain, particularly affecting fish, amphibians and some large pond invertebrates.

Crassula helmsii (*see photo 1 below*) has shoots that carry pairs of parallel-sided leaves, each between 4mm and 23mm in length. In summer, small white flowers, each with four petals, emerge on long stalks from out of the water.

Parrot's feather (Myriophyllum aquaticum)

This South American plant (*see photo 2 below*) first got into the wild in 1960 after accidentally spreading from garden collections. It grows well in nutrient-rich waters, so much so that public waterways are regularly blocked by this species in particular.

The emergent leaves of this pond plant are stiff, while submerged leaves are more fragile. Leaves come from the stem in whorls of between four and six. As the name suggests, the protruding stems and leaves look almost feather-like from a distance when poking out of the water.

Water fern (Azolla filiculoides)

This floating aquatic fern (*see photo 3 below*) found its way into the wild after featuring heavily in ornamental collections. It outcompetes native species by forming dense mats so thick that animals and humans can fall into the water, failing to see the water surface. These mats are made up of thousands of individual plants, each 1–2cm across, with green leaves tipped with orange or red at the edges.

Floating pennywort (Hydrocotyle ranunculoides)

Floating pennywort used to be sold under the misleading name marsh pennywort, which is a native species. It is commonly found in the wild in the southeast of England and is steadily

making its way further north and west, including into Wales. This species can grow up to 20cm in a single day. It has fleshy stems and rounded 'toothed' leaf edges.

BE VIGILANT ABOUT YOUR POND Because of the threat that these invasive problem plants bring to the wider countryside, a government ruling has now banned their sale in the UK. As of 2014, if any garden centre, nursery or other horticultural retailer is caught selling any of these species (as well as Water primrose (*Ludwigia grandiflora*, *Ludwigia peploides* and *Ludwigia uruguayensis*), they may face a hefty £5,000 fine and even time in jail.

Many scientists and wildlife conservationists are hopeful that this ban will have a dramatic impact on the spread of invasive pond plants in the UK. However, this change in the law will not eliminate the problem of invasive pond plants entirely because they are already reproducing in the wild, as well as in many of our urban and suburban ponds. These escapees will continue to be spread easily and widely between existing ponds through the movement of cuttings as well as through pond-dipping nets, buckets and even wellies. You can reduce the accidental spread of invasive pond plants by carefully washing your pond equipment if moving from pond to pond and ensuring that you carefully clean cuttings of exotic stowaways before you add them to your pond.

MANAGING AND REMOVING INVASIVE POND PLANTS If you notice any of these invasive plants in your pond, the good news is that many of them can be managed through regular netting or raking across the surface of the water. In larger ponds, you can use a floating boom to sweep the surface waters. The bad news is that this may not be a battle you end up winning in the long term. This is because many invasive pond plants can regrow from fragments, so it helps to be careful how you go about removing plants. For instance, a small patch of Australian swamp stonecrop on one side of the pond can spread quickly to other parts of the pond if shredded into tiny pieces by an aggressive management session with a rake.

Some invasive non-native plant species can be managed by the addition of shade. A strategically planted tree or shrub can slow the growth of some exotic pond plants, though this brings its own problems by way of falling leaves. In the past, some pond owners have taken the extreme step of using black plastic to smother problem pond plants out, starving them of light for a season or two. Other pond owners have been forced to apply pesticides. If you are considering this route, you will need to talk to a qualified freshwater specialist.

POND ANIMALS TO LOOK OUT FOR

ONE OF THE GREAT JOYS of ponds is that every day brings with it new encounters with a whole range of unusual animals, many of which can be immensely satisfying. These include bats, dragonflies, grass snakes and newts but also far smaller and overlooked creatures like springtails, water fleas and water mites, for instance.

The other great joy of ponds is that, even in the deep midwinter when the landscape looks sparse and lifeless, there is plenty of aquatic activity under the surface for you to observe and study. In a given year, many ponds will be home to 100 or more species observable with the naked eye. It really is all in the looking.

In this chapter you will find details of the classic pond animals you might see in or around your wildlife pond throughout the year. As well as detailing the specific species to look out for, this chapter also includes information about how these animals come to colonise wildlife ponds and how soon you might come to expect them.

MICROSCOPIC COLONISERS

Without a shadow of a doubt, some of the most monstrous and bizarre animals to call your pond home are those that are almost indiscernible to the naked eye. Many of these tiny organisms reside on the undersides of submerged pond plants and can be carefully teased off and put on a microscope slide with a drop of water. They include some of the most incredible and hard-wearing animals to ever have lived, including water-bears (more properly called tardigrades) and (my personal favourite) the bdelloid rotifers, a group of animals that haven't had sex for perhaps 80 million years. Part of the thrill in looking out for these creatures is celebrating the even tinier organisms that fizz and whir across your magnified field-of-view whilst searching, including protozoans such as *Amoeba* and *Paramecium*.

How will they arrive?

The short answer is: quickly. Many of these animals move from water-source to water-source by dehydrating their bodies (or eggs) into a tiny husk, capable of blowing from place to place on prevailing wins, often for many years. Once these dried-up husks land in a puddle or pool, within hours they hydrate and – hey presto! – they are ready to multiple once more.

ROTIFERS

For many decades, early microscope-wielding scientists thought these 'animalcules' possessed spinning wheels upon their upper parts. In fact, these 'wheels' are actually tiny hairs (called cilia) that beat in a rhythmic manner to create a tiny cyclone of current that pulls in passing food particles. Among the first visitors to your pond is likely to be a bdelloid rotifer called *Rotaria*, which occasionally reaches a millimetre or more in length though is usually half this. Highly active, it moves from place to place rather like a tiny leech whilst searching for food-rich waters.

RIGHT **1** Bdelloids rotifers are among the first arrivals to new ponds. **2** Tardigrades often congregate upon mosses at the water's edge. **3** A hydra dispatches a water flea.

TARDIGRADES

Tardigrades (or water-bears as they are often known) look almost like a poorly imagined balloon-animal. Their gently plodding manner and their stout body gives them an unrivalled charisma compared to other microscopic pond animals, making them a firm favourite of Victorian microscopists. Though many tardigrade species favour damp places like wet moss and roof gutters, some species (including those of the genus *Macrobiotus*) are truly aquatic.

HYDRA

With cousins that include the jellyfish and corals, hydra have their evolutionary roots in one of the most primitive yet productive parts of the animal kingdom. From a jelly-like stalk tiny tentacles grab at passing food particles which are slowly digested. Some individuals reach a centimetre or more in length and can be seen with the naked eye, dangling from rocks or the leaves of pond plants at the pond edge. The green colouration of one common species (*Hydra virissima*) comes from colonies of photosynthesising single-celled algae called *Chlorella* that it keeps within its body.

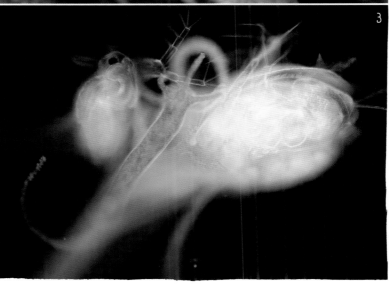

WORMS AND WORM-LIKE CREATURES

Worms and worm-like creatures are important and often overlooked members of your pond community. Some graze the algae, keeping the pond clean. Others burrow in the substrate. Many are predators that actively hunt other prey. They include some well-known groups such as leeches, which are among the most misrepresented of UK invertebrates. Among the other worm-like animals in ponds are roundworms, horsehair worms and ribbon worms. Each of these worm-like creatures is likely to turn up in your pond at some point so watch out for them, and when you see them, allow yourself a moment to marvel at their simple yet very effective way of life.

How will they arrive?

Some worms and worm-like creatures will move through soil to find your pond but most will arrive after hitching a ride on a range of hosts, including birds, snails and amphibians. Others are likely to come upon your pond via eggs (some of which will blow into the water). Many of the species included on these pages are likely to find your pond within a matter of weeks.

FLATWORMS Free-living flatworms (which differ from their parasitic cousins, the flukes and tapeworms) will be a common sight in nearly all ponds. They glide across the edges and the bottom of the pond upon rows of tiny beating hairs between which flows a trail of mucus. Flatworms are broad and flat and have a noticeable set of eyes, which give them a certain charisma not present in most invertebrates.

Flatworms have interested scientists for centuries because they represent some of the simplest and most easy-to-study experimental animals. These tiny animals are capable of responding to light and dark, they are capable of fleeing from predators and they have incredibly interesting sex lives. Almost like an amoeba, flatworms often choose to reproduce by splitting in half. When the time is right, they set adrift their tail, which moves off on its own before growing a new head.

Britain is home to approximately 60 freshwater species of flatworms and you can find many of these by carefully turning over stones under the water or looking through the pond's decomposing leaves.

Perhaps the most common flatworm you might encounter in your wildlife pond are species of the genus *Dugesia*. Commonly, *Dugesia* reach a size of about 20mm. Each of the species in this genus possesses two eyespots and has a rounded, arrow-like head.

Another flatworm common to many ponds is the white flatworm (*Dendrocoelum lacteum*). The semi-transparent nature of this species means that, with a hand lens, one can observe the sight of its greenish food paste moving rhythmically down into the gut.

SPINY-HEADED WORMS

Many of your pond's worm-like creatures will be parasites that live within the bodies of other creatures. This makes them very hard to spot. They include the numerous species of tapeworms and

RIGHT Eider ducks are one of many waterfowl species that act as a final host for parasitic spiny-headed worms. A single bird may carry as many as 750 worms.

flukes often found within fish. One parasitic worm you may be able to spot is *Polymorphus* – a type of so-called spiny-headed worm. In its larval form, this parasite lives within the body of the freshwater shrimp. It can be spotted as a bright orange blob in the middle of the shrimp's body. This colourful blob is likely to serve as a cue that catches the attention of hungry waterfowl. These waterfowl gleefully swallow the freshwater shrimp, not knowing that they have unintentionally become a host for the spiny-headed worm's adult life stage.

LEECHES When most people consider leeches they imagine the famous medicinal leech (*Hirudo medicinalis*), which can grow up to 8cm long. The medicinal leech is the only leech in the British Isles capable of biting through the skin of large mammals (including humans), although amphibians and fish offer easier opportunities in most ponds. Sadly (depending on your point of view) this species is nearly extinct in the UK, mostly because of land drainage but also because so many of these leeches were caught in previous centuries for medicinal use.

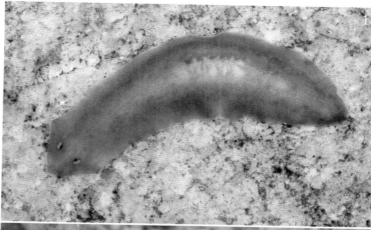

ABOVE **1** Flatworms have notable eye spots that you can see with a magnifier. **2** Looking down on a fish with fish leech clearly visible.

Your pond may play host to other (no less interesting) leech species, which can all be diagnosed through their 'looping' means of travel or via their charismatic way of swimming like a ribbon undulating through the water. If you catch leeches in a net and transfer them into a pond-dipping tray, you may need to leave them for a minute or two so that they can change from a defensive globular shape into a more recognisable and lively form.

One leech species common to many ponds is *Hemiclepsis marginata*, the fish leech. Though fairly innocuous to look at (adult size is 20mm), this species can become problematic for some ornamental fishkeepers. The fish leech feeds mostly around the mouth and gills of fish, and sometimes multiple leeches attach themselves to a single individual. I have seen them many times feed upon amphibians, often near their armpits and groin areas. Indeed, so prolific are these creatures that I have even seen them attach to tadpoles and froglets, perhaps to the confusion of all involved.

A common leech you may spot are leeches of the genus *Erpobdella*. These leeches rarely grow to a length of more than 40mm and are common and widespread throughout the UK. *Erpobdella* are active predators. They ingest a range of prey that includes insect larvae and worms.

FLIES AND THEIR LARVAE

Like beetles, flies are an order of insect (Order Diptera) that the world could not do for long without. Flies energise food chains with their industrious nature and their incredible diversity. In all, the world may be home to more than 1 million species of fly, and many of these species use freshwaters as a nursery ground for their larval life stages, so you should look out for them and cherish each and every one.

Though some flies are reviled for their ability to carry diseases, particularly midges and mosquitoes, nearly all flies are of no threat to us at all. In fact, popular to contrary belief, only a handful of midge species bite – most are harmless pollinators. For this reason, spending time quietly observing the larval development of flies in ponds is an excellent way to get a feel for how flies have become such a success in the history of life on Earth.

MOSQUITOES In the UK, mosquitoes split neatly into two groups. The larvae of Anopheline mosquitoes rest horizontally submerged just under the surface of the water. When disturbed, these mosquito larvae swim with a jerky motion. The larvae of the other mosquito clan (the Culicines) hang downwards into the water with only their bottoms (through which they breathe) protruding through the surface tension of the water. In both groups, their pupae retain the ability to swim. When threatened, the pupae can activate a comedic spinning 'body roll', which helps them avoid predators.

Of the two groups, the most annoying mosquitoes for humans are the Culicines, particularly *Culiseta annulata*, a large mosquito with spotted wings, which commonly breeds in water butts. Though the thought of attracting such creatures to your pond may sound troubling, remember that most of the larvae and adults will become prey for something else, be it newts, frogs, bats or birds. Plus, remember that only the females feed upon blood. Male mosquitoes are yet another important pollinator.

MIDGES Most of the general public despise midges almost as much as they do mosquitoes, but in truth, the reputation of midges is wholly unjust. This large group of flies includes the families of the phantom midges, owl midges, gall midges and meniscus midges, all of which have little interest in humans. Only a single midge family (the Ceratopogonidae) can bite – the most noteworthy being the highland midge (*Culicoides impunctatus*).

How will they arrive?

Flies are likely to be one of the first visitors to your pond. On a sunny day, you will see them laying their eggs at the pond edges or dancing across the pond surface, dropping them into the water. If you see them doing this you should rejoice! Many of these eggs will hatch into algae-eating larvae, some of which larger pond animals will hunt. Congratulations, your pond's food web is kicking into action!

One of the earliest visitors to your pond is likely to be one of a few species of chironomid flies, often called non-biting midges. Their eggs hatch into distinctive red larvae called bloodworms, which live in the water. Bloodworms are adapted to live in low-oxygen environments and therefore possess internal body fluid rich in iron, which gives them a distinctive red colour. These larvae can sometimes reach a body length of 25mm and they are incredibly important food for a host of species.

In late spring, some ponds become home to the strange, ghost-like larvae of the phantom midges. These horizontal swimming larvae (10mm or so in length) are almost totally transparent, apart from the internal organs within and the jaw-like antennae used for catching prey. Mostly, phantom midge larvae give away their presence in the water through sudden jerky swimming movements.

HOVERFLIES Some small ponds and smelly pools may be home to the larvae of *Eristalis* species, one of which is the dronefly (*Eristalis tenax*) – a common hoverfly that looks a bit like a honey bee drone. The larvae of these hoverflies are often called 'rat-tailed maggots' for obvious reasons. The long tail, which occasionally reaches 150mm in length, is actually a tube through which the larvae breathe.

CRANE FLIES Many species of crane fly begin life as pond-dwelling larvae. These early life stages possess a pair of spiracles on the rear end, surrounded by finger-like projections through which the larvae breathe. When the time comes, the larvae move out of the water to pupate. The larvae of some crane fly species are incredibly camouflaged. For instance, *Phalacrocera replicata* has a green body covered in filamentous hair-like structures, which helps it hide among the submerged moss upon which it feeds.

RIGHT **1** Bloodworms (midge larvae) can be very common in ponds. **2** The rat-tailed maggot, a charismatic larvae of the dronefly. **3** Many cranefly species use ponds as a nursery ground for their larvae.

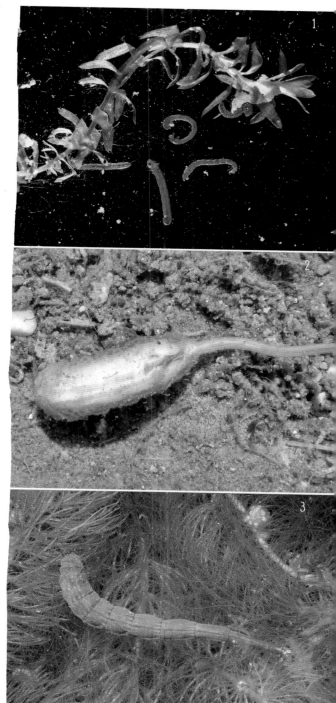

WATER BEETLES

Britain's freshwaters are home to a dazzling variety of water beetles and their larvae. In total, around 350 species can be found and each beetle has adapted to a particular niche or zone within their freshwater environment. In ponds, there are tiny species like *Haliplus fulvus*, which crawl along water weeds. There are surface-dwellers like the whirligig beetle (*Gyrinus* spp.) and there are scavengers such as the small silver beetle (*Hydrochara caraboides*). And then there are the celebrated diving beetles, a family of often large and dynamic water beetles famed among pond watchers for their athletic prowess and carnivorous habits. The larvae of some diving beetle species, particularly those of the great diving beetle (*Dytiscus marginalis*), are actually more fearsome-looking than the adults.

Water beetles breathe by pulling a bubble of air under their wing cases, which they use a bit like a scuba tank during their frequent dives. Their presence in ponds is often given away by their frequent returns to the surface to top up these air supplies.

WHIRLIGIG BEETLES Whirligig beetles possess eyes that are split across the middle – the lower half is for seeing underwater, the upper half is for scanning for threats above the water surface. These medium-sized water beetles are known for their habit of skating across the surface of the water. Sometimes, vast clouds of whirligig beetles zigzag and circle one another in the middle of a pond, giving off rippling patterns that, to the human observer, have a certain captivating intensity to them. To keep from bumping into one another during these speed-skating frenzies, whirligigs sense for reflected pressure waves in the water. As such, they are one of the world's only echolocating insects.

BELOW Male and female great diving beetles can be told apart by their wing-cases. Females have highly striated bodies to stop over-eager males from gripping them too tightly during the breeding season.

How will they arrive?

Nearly all water beetles are 'flight ready' as adults, but most species will only search out ponds when the conditions are exactly right. Water beetles can turn up in moth traps at night, suggesting that at least some species fly at night, perhaps using the moon as a navigational aid. For new ponds, water beetles will be among the first animals to arrive. On warm, windless days in midsummer, I have seen great diving beetles turn up in ponds within just three days of them being filled.

DIVING BEETLES In the UK, there are six species of diving beetle. The most widespread is also the largest – the 30mm-long great diving beetle (*Dytiscus marginalis*). This species swims with powerful hind legs, seeking out tadpoles and large invertebrates that it pulls apart with its impressive jaws. Males differ from females in possessing shiny, smooth wing cases that differ from the female's deeply grooved wing case patterns. Males grasp on to females with a pair of front feet specially equipped with strong suckers.

All adult diving beetles can be active throughout most of the year, though many will enter a deep winter slumber during periods of icy weather. The peak time to see adult diving beetles is late summer and in early autumn. Late spring is a great time to see their monstrous larvae, some of which can reach a length of 5cm or 6cm. These larvae move by running with paddle-like legs through the water. They hunt and kill prey using a pair of broad and sensitive eyes, and impressive scythe-like mandibles. So voracious in their feeding habits are the larvae of the great diving beetle that I have seen them chase away fish more than triple their size! They also regularly hunt tadpoles.

Some pond owners become a little unnerved having such a large invertebrate predator (and its carnivorous larvae) within their ponds, but fear not, for these water beetles are very much the natural order of things. Their presence is one of the reasons that frogs and toads have adapted to produce so many offspring.

SILVER WATER BEETLES In many ways, the silver water beetles resemble the diving beetles, but they differ in their club-shaped antennae and their hairier underside – an adaptation that allows them to trap an extra layer of air as the beetle dives downwards through the water. Unlike diving beetles, adult silver water beetles are mostly scavengers or herbivores.

The larvae of the great silver water beetle (*Hydrophilus piceus*), some of which can grow to an impressive 7cm in length, feed upon freshwater snails, drilling holes into their shells to get to the nutritious insides. Upon reaching large size, the larvae pupate in the mud at the bottom of the pond. Adults can live for three years or so, though many will die in their first year of life.

In Victorian times, the great silver beetle became something of a celebrity among parlour aquariums, leading to a bustling beetle trade emerging in pet shops and aquarists. So high did the commercial value of the great silver beetle rise that London's ditches were said to be cleared of them, courtesy of amateur collectors eager to make a quick profit. Today, the species has a patchy distribution across the UK but it does occasionally turn up in some garden ponds.

BELOW The monster-like larvae of the silver diving beetle are the stuff of nightmares for watersnails.

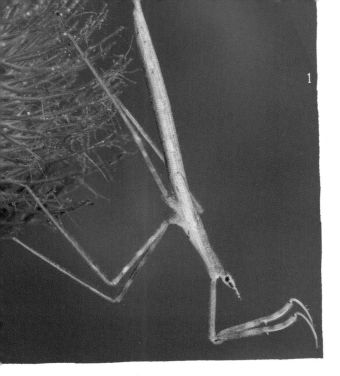
1

WATER BUGS

- -

To small fish and tadpoles, most water bugs are the stuff of nightmares. Grasping prey with their front legs, these agile predators stab syringe-like mouthparts into their captive before sucking up their bodily juices. The most common of water bugs are also the most impressive fliers – the backswimmers and the saucer bugs – but there are other species that look almost as if they belong in the tropics. The water scorpion (*Nepa cinerea*) and the water stick insect (*Ranatra linearis*), for instance, have startlingly exotic demeanours and their assassin-like hunting behaviour really is something to behold. But not all water bugs are monstrous predators. Some, like many species of lesser water boatmen, are entirely herbivorous. They include among their ranks the noisiest animal for its size in the world, *Micronecta scholtzi*.

BACKSWIMMERS Also known as the greater water boatman, the common backswimmer (*Notonecta glauca*) is a widespread predator of tadpoles, invertebrates and even small fish. By swimming upside down, the backswimmer can scan the pond bottom for possible prey. Backswimmers are particularly drawn to vibrations given off by drowning insects on the surface. Their mouthparts allow them to deliver a toxic bite, so handle them with care.

WATER BOATMEN The UK is home to almost 40 water boatmen species. Like backswimmers, these insects dash through the water via a pair of powerful oar-like legs. Being smaller, they are more capable fliers than backswimmers, meaning that they can appear in a range of pond-like habitats, including water butts and cattle troughs. These are highly vocal pond creatures. Like grasshoppers, males sing (although stridulate is a better word) to attract attention at dusk. You will need to get close to the pond surface to hear this noise. *Micronecta scholtzi* is the record-holder – you may hear this 2–3mm-long water boatman from 5m away!

THE SAUCER BUG The saucer bug (*Ilyocoris cimicoides*) looks like a giant swimming bedbug. It seeks out tadpoles and water mites, which it can dispatch with apparent ease.

How will they arrive?

- -

Most water bugs are impressive fliers, but some (particularly the water stick insect) do so rather clumsily. When the moment takes them, these pond invertebrates come to water's edge, and fastidiously clean their wings, eyes, mouthparts and erect wings hidden under tough wing cases. Many water bugs will find their way to new ponds within the first year depending on the food sources available to them. When ponds start to dry up during the summer months, many water bugs will move on to larger ponds where more wholesome feeding opportunities may exist.

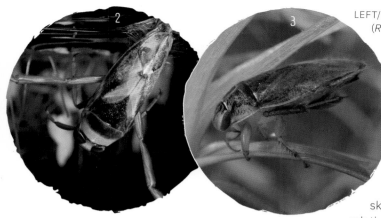

LEFT/OPPOSITE/BELOW **1** A water stick insect (*Ranatra linearis*) on the lookout for prey. **2** A common backswimmer. **3** A saucer bug. **4** A common pond skater.

POND SKATERS AND OTHER SURFACE BUGS

The common pond skater (*Gerris lacustris*) is a predator familiar to the surface of many ponds. The tips of its legs are covered in water-repellent hairs, meaning that this predator can skate atop the water's surface tension with relative ease. Pond skaters are attracted to the vibrations of terrestrial insects drowning in water, which they speedily dispatch with their pointed mouthparts. Another surface bug common to some ponds is the water measurer (*Hydrometra stagnorum*). This rather dainty predator prefers to spear creatures moving just under the water surface rather than on top.

At night, the pond surface is prowled by a different group of surface-dwelling bugs. These are the so-called 'water crickets'. The most common species is *Velia caprai*, which measures up to 8.5mm.

Sometimes, almost as if driven by instincts beyond their control, saucer bugs kill and don't even feed upon the spoils, leaving them for something else to scavenge. Like backswimmers, these are the sharks of the water bug tribe, and they can pack a powerful bite if handled roughly.

WATER STICK INSECTS Larger ponds are sometimes home to a bug that has evolved a stick-like shape for camouflage. This is the water stick insect, a widespread but patchily distributed bug that you may have the fortune of seeing in your pond if it happens to be big and weedy enough. Water stick insects are ambush predators. They hide among leaves, lashing at passing insects and tadpoles, which they grasp and consume with pointed mouthparts.

WATER SCORPIONS

Just as the water stick insect isn't really a stick insect, the water scorpion is not really a scorpion. Rather, these water bugs have a long pointed 'tail', through which they breathe, and stout 'claws' that they use to grab at passing prey. Like water stick insects, these animals are rarely spotted without the aid of a pond-dipping net and inspection tray. Even then, it is easy to mistake water scorpions for leaves.

POND CRUSTACEANS

There are 67,000 crustaceans on Earth, and many species have adapted to colonise and live within freshwaters, including in ponds. Like insects, crustaceans must shed their skin as they grow, but they differ in the number of limbs they possess and in their bizarre larval forms, many of which are zooplanktonic.

Of the crustaceans in your pond, the most important to the healthy functioning of its ecosystem are the tiny water fleas. These animals bloom in their thousands during parts of the year, providing food for a host of invertebrates, fish and newts. Other crustaceans, including freshwater shrimps and water hog-lice, play an equally important role by decomposing pond detritus. There are also some easy-to-miss crustaceans in ponds, including the lesser-known seed shrimps and their tiny crustacean cousins, the copepods.

WATER FLEAS The proverbial name 'water flea' comes from the small size and jerking movements of this tiny but incredibly important

pond crustacean. Eighty species are known to Britain (the most famous being species of the genus *Daphnia*), although more species of water fleas are undoubtedly lurking out there completely undiscovered. Though they look little more than dots to the naked eye, under a microscope one can see a host of interesting features, including a beating heart and pumping gut, and the familiar crustacean antennae, which most water fleas use for swimming. Water fleas like *Daphnia* have an impressive ability to produce clonal offspring, meaning that ponds can exponentially fill with water fleas within a matter of days. This is good news for most pond owners because water fleas feed upon microscopic algae and bacteria in the water. The combined actions of these tiny creatures can make pond water go from cloudy to clear within days. These blooms normally happen in late spring and then again in autumn, when a sexual batch of offspring is produced, from which come the hardier eggs that will see out winter.

COPEPODS Like water fleas, planktonic copepods (shown above) are sometimes blown in on the wind. They can be common to many ponds and puddles, even in water-logged tyre tracks and stagnant pools in the holes of old trees. With two long antennae, noticeable paired egg sacs attached to either side of the body and a single light-sensitive eye, the most common are species of the aptly named genus *Cyclops*.

SEED SHRIMPS At first glance it is easy to confuse seed shrimps with water fleas. Under a microscope, however, a world of difference occurs. Like miniature clams, seed shrimps are covered in a pair of tiny plates. Their legs dance in and out of these plates to assist with movement and feeding, giving them a comically secretive appearance. Seed shrimps (*Cypris* spp.) are ostracods, a branch of the crustacean family tree seemingly unchanged since their appearance in the fossil record some 500 million years ago.

How will they arrive?

Lacking the ability to fly, many crustaceans are slow to colonise garden ponds. Most are probably brought in as eggs stuck to birds, particularly waterfowl, which then hatch and flourish in new ponds. The eggs of some crustaceans, such as water fleas, can be incredibly hardy, and most can withstand almost total desiccation. During dry periods, these eggs sacs are blown throughout the landscape. If they land in a pond or large puddle, a new generation will flourish, sometimes within days, weeks or months.

FRESHWATER SHRIMPS The freshwater shrimp species found in most garden ponds is *Crangonyx pseudogracilis*, a 10mm freshwater crustacean that moves upright (rather than on its side) through detritus at the bottom of the pond. This common freshwater shrimp is actually a non-native species accidentally introduced from North America in the last century. Thankfully, it does not appear to have caused harm to native species.

WATER HOG-LICE These woodlouse-like crustaceans are so-named because they resemble the parasitic lice that live on pigs.

ABOVE **1** Freshwater shrimp thrive in detritus-filled ponds. **2** Water fleas. **3** A tiny seed shrimp. **4** A rather dusty water hog-louse.

Like a woodlouse, the water hog-louse (*Asellus aquaticus*) walks upon seven pairs of legs, searching out decaying leaves upon which it feeds. Hidden pairs of modified legs situated underneath the body provide for a special set of gills, which can be seen pumping energetically when inspected with a magnifying glass. These gills allow the water hog-louse to live in low-oxygen environments, including the most stagnant and oxygen-deprived ponds.

POND SNAILS

Ponds are home to a range of freshwater snails, each of which does an admirable job of consuming algae that grows upon rocks and the surface of pond plants, although some species also graze the pond plants themselves. Many popular pond guides consider pond snails a 'must-have' animal to keep ponds clean, but the truth is that pond snails tend to recycle nutrients within the water rather than remove them entirely. Instead, the ecological value that freshwater snails bring to ponds is through their role in food webs. Many birds, such as blackbirds (*Turdus merula*) and starlings (*Sturnus vulgaris*), eagerly dip their beaks into water to feed upon pond snails. Other predators, including some water beetles, depend upon snails as a food source for their larvae, and some animals, including many worms and worm-like animals, use them parasitically as life-giving hosts.

Of the 40 or so freshwater snails in Britain some, like the 3mm-wide nautilus ramshorn (*Gyraulus crista*), are truly tiny. The giant of freshwater snails is the great pond snail (*Limnea*

BELOW **1** The great pond snail – note the chunky antennae. **2** Ramshorn snails get their name from their notably flat shells. **3** The wandering snail.

How will they arrive?

Most freshwater snails arrive in new ponds through the addition of pond plants, many of which will have slimy snail eggs hidden upon them. In the wild, pond snails are likely to move around to new ponds by hitching a lift on birds' feathers, particularly those of ducks and other wildfowl. Interestingly, research suggests that some terrestrial snails can survive the perilous journey through a bird's digestive system. After the bird swallows them, the snails come out alive and well in their droppings. It may turn out to be that some freshwater snails can manage the same feat and use this technique as a handy way to colonise new ponds.

stagnalis), which can grow to almost 4cm. This snail is common to most wildlife ponds.

Freshwater snails occur in a range of pond habitats though they can be limited by the amount of calcium available. This can mean that in particularly acidic ponds, snail populations may be dramatically reduced or even entirely absent.

In this section, we feature three of the most common snails you are likely to see in your pond. Each freshwater snail species lays eggs in strips or blobs of translucent jelly applied to the pond substrate, to rocks or to the undersides of pond plants. Under a microscope, you can see the tiny shells of the embryos as they develop within this life-giving jelly.

GREAT POND SNAIL
The great pond snail (*Limnea stagnalis*) is a freshwater champion. Growing to almost 5cm long, this large and widespread snail has a distinctive cone-shaped shell with a wide aperture, from which the snail's muscular foot and head sprouts. The species lays its eggs in sausage-like deposits of jelly, which are often found on the undersides of water lily leaves.

RAMSHORN SNAILS
This charismatic clutch of freshwater snails each has a shell that is coiled into a disc shape, held high above the foot of the snail. The two species most likely to find their way to your pond are the great ramshorn snail (*Planorbis corneus*), with a 30–35mm diameter shell, and *Planorbis planorbis* (18mm diameter), known simply as 'The Ramshorn' in many mollusc identification books. Both species of freshwater snail have within them distinctive red fluid that looks like blood. This gives their muscular foot an orange tone, particularly observable in newly hatched ramshorn snails.

WANDERING SNAIL
The wandering snail (*Radix peregra*) has a coiled shell that is more streamlined and held more closely to the body than other common freshwater snails. This shell

Pea mussels

Another molluscan visitor to some ponds is the tiny clam called the pea mussel, many of which rarely exceed a length of 5mm or so. Pond dippers can easily miss pea mussels because most lay buried among the pond sediment. Their presence is often given away by the empty valves that wash onto the pond shore, sometimes in great numbers. Some species of caddisfly nymph use these discarded valves in their cases.

As with other pond creatures, mussels likely travel from pond to pond via birds but, interestingly, amphibians might play a part too. Reports of freshwater mussels attaching themselves to the tails and legs of newts have been known, which could provide some mussels with a viable, if clunky, means of terrestrial transport to new ponds.

regularly reaches a height of 20–30mm. The term 'wandering snail' probably refers to the fact that these are very active molluscs and rarely seem to stop moving around the pond. They are also very good colonisers, and may be among the first snails to turn up in your pond.

MITES AND SPIDERS

In this section, we consider the arachnids – specifically, the water mites and spiders.

One of the most interesting and easily overlooked creatures in your pond will be the water mites. Most water mite species come in a range of colours and their species exhibit a number of different lifestyles. Many species move through the water by frantically pumping their legs in a manner that can best be described as cartoonish. As a group, the water mites are drastically under-recorded. There may be hundreds of species awaiting discovery.

Ponds offer slightly fewer opportunities for spiders, although a number of spider species have adapted to make freshwaters their own in various ways. Almost all of these species are surface or pond-edge species, though one – the water spider (*Argyroneta aquatica*) – has adapted against all odds to a life under the surface. Of 45,000 spider species known to scientists, the water spider is the only species able to achieve this.

WATER MITES (*pictured below*) Europe is home to almost 1,000 species of water mite, each of which can be highly variable in colour. Some species are a striking blue, others come in dashing yellow forms and many are an eye-catching red, each probably informs would-be predators that the mites are not worth swallowing. Among the most rotund water mites in large ponds are species of the genus *Eylais*, some of which reach a diameter of 6mm or so across, but most water mite species are not more than 2mm in size.

PIRATE WOLF SPIDER The pirate wolf spider (*Pirata piraticus*) has a body length of between 5mm and 9mm, and hunts insects at the water's edge. This fast and very active wetland spider has water-repellent hair on its legs, which allows it to move across the water surface if threatened by predators. It can be easily identified by the short, yellow-brown V-marking to the rear of its abdomen.

STRETCH SPIDERS

STRETCH SPIDERS The 'stretch spiders' get their name from their habit of resting with their long legs stretched out in front of the body, providing a stick-like means of camouflage. One species of stretch spider, *Tetrognatha extensa*, is notably common near still freshwaters, including ponds. This spider has a body between 6.5mm and 11cm long.

RAFT SPIDERS Raft spiders are so-called because of their ability to run on water while fleeing predators and also when hunting. There

are two species of these semi-aquatic spiders in the UK. The largest is the fen raft spider (*Dolomedes plantarius*), whose females regularly reach a body length of 22mm with a leg span of up to 70mm. Sadly, raft spiders are incredibly rare in the UK and they are highly unlikely to colonise garden ponds.

WATER SPIDER The water spider (*Argyroneta aquatica*) maintains an aquatic lifestyle through special hairs that trap a layer of air between its body and the surrounding water. This adaptation allows the spider to effectively 'breathe' underwater. Water spiders also use this silky layer of hair to carry air from the surface of the pond to underwater webs that resemble a diving bell. These dome-shaped webs are used as an underwater safe house, in which baby spiderlings can be raised. Interestingly, oxygen in the water diffuses naturally through this webbing as if acting like a huge lung. Though widespread, water spiders are patchily distributed across the UK and Ireland, and are not likely to colonise new ponds easily. They prefer large weedy ponds that provide the architecture required for their incredible webs.

How will they arrive?

Most water mites will hitch a lift to ponds via large pond insects, upon which they act as external parasites while in their juvenile life stages. Host species regularly include backswimmers and water beetles. Some spiders, including wolf spiders, may 'balloon' into new habitats when they are juveniles. 'Ballooning' involves spiders releasing skywards a long string of gossamer-like thread that catches the wind and sends them soaring upwards. Juvenile Pirate Wolf Spider (*Pirata piraticus*) may travel for miles using this method.

DRAGONFLIES AND DAMSELFLIES

One of the biggest compliments that nature can pay to you and your pond is the sight of dragonflies and damselflies zipping and zooming over the surface of the water. These industrious predators make short work of midges and mosquitoes around the edge of the pond, and their underwater life stages (called nymphs) are one of the pond's most capable underwater predators.

On the whole, dragonflies differ from damselflies in the way that they hold their wings at rest. Damselflies have long abdomens and hold their wings closed against their body, while dragonflies hold their wings out horizontally from

How will they arrive?

If your pond is not too isolated, dragonflies and damselflies often arrive within the first year, drawn to the pond to feed upon flies and other insects hatching out of, or laying eggs in, the pond. Clean ponds of small or medium size often have damselfly nymphs in them within the first two years.

the body when perching. Dragonflies are very agile and swift while flying, whereas damselflies are often smaller and daintier in the way that they fly.

Damselfly and dragonfly nymphs are superficially very similar. Upon their underside is a distinctive 'mask' – a set of jaws that they can launch at prey should it stray too close. Young damselfly and dragonfly nymphs spend much of their time going after bloodworms and water fleas, but the prey they favour changes as they grow. Large dragonfly nymphs regularly take small fish and tadpoles. Indeed, their presence in large numbers can have a dramatic impact on your pond community.

Dragonflies can be neatly separated into three groups according to their flying styles. There are the hawkers, the chasers and the darters. Hawkers are best known for patrolling their territory in very showy fashion. Chaser dragonflies are highly territorial, with males regularly scuffling for possession of a given pond. Darters spend much of their time perched, waiting for passing insects, which they dart out towards to capture and feed upon.

DAMSELFLIES In many ponds, the first damselfly to colonise your pond will be the blue-tailed damselfly (*Ischnura elegans*). Blue-tailed damselflies can be quite variable in their colour

LEFT/OPPOSITE **1** Unlike other damselflies, the Emerald Dragonfly (*Lestes sponsa*) often perches with wings half-open. **2** Broad-bodied chaser. **3** Emperor Dragonfly (*Anax imperator*) – see also picture on page 100.

patterns, but males have an electric blue segment near the end of their tails that is easy to spot when it is resting. Another common coloniser of new ponds is the large red damselfly (*Pyrrhosoma nymphula*), which can be seen in even the smallest of ponds. Both species are on the wing between late April and September.

A slightly less common, but no less attractive, damselfly is the emerald damselfly (*Lestes sponsa*), which has an attractive metallic green colour to its abdomen. Males have distinctive blue eyes, and adults of both sexes fly from late June to September.

HAWKER DRAGONFLIES

The common hawker (*Aeshna juncea*) can be a regular and widespread visitor to ponds across the UK. Both males and females have bronze-coloured leading edges to their wings. Adult males have blue eyes and blue markings with dashes of yellow along the abdomen. Adults fly between late June and on sunny days as late as October.

Another dragonfly likely to visit your pond is the impressive emperor dragonfly (*Anax imperator*). This dragonfly has a wingspan up to 10cm and adults are often seen patrolling ponds in warm sunshine at dusk. This species can be particularly common to ponds in southeast England.

In some parts of the UK, particularly in the south and west, your pond may receive occasional visits from the golden-ringed dragonfly (*Cordulegaster boltonii*). This dazzling yellow and black dragonfly is one of our largest. It flies between June and August.

CHASERS AND DARTERS

Of the chasers and darters, the four-spotted chaser (*Libellula quadrimaculata*) is the largest and bulkiest. Both males and females of this species have two spots on each wing and their hind wings have a black patch at the base. These dragonflies can be very common except in the north of England and Scotland. The four-spotted chaser favours shallow ponds with plenty of emergent vegetation.

Like the four-spotted chaser, the broad-bodied chaser (*Libellula depressa*) is also unlikely to be found too far north. Both sexes have distinctly flattened abdomens, and breeding males are blue. This species is often one of the first dragonflies to colonise new ponds.

Another common visitor to most ponds is the common darter (*Sympetrum striolatum*). This dragonfly often flies far from water, so can be quite quick to colonise isolated ponds. Even small ponds are of interest to this species. The common darter is one of the latest dragonflies on the wing in the UK; adults can sometimes still be on the wing in November.

CADDISFLIES, MAYFLIES AND MOTHS

Scientists use different names to describe the life stages of different insects. In caddisflies, dragonflies, damselflies, mayflies and stoneflies, the early life stage is called a nymph rather than a larva. Unlike most larvae, nymphs resemble the adult form as they grow (albeit lacking wings) and they emerge into adulthood via a final moult rather than undergoing a pupal stage. In late spring and summer, this moulted skin can be seen dangling off plants around the pond edge or simply discarded on the water surface. The adult life stage of these insects rarely lives long. Famously, many species of mayfly may only live for a matter of days or weeks as adults. The same is also true of caddisflies.

Caddisflies mostly fly at dusk or during the night, meaning that the adult life stages of these species can be hard to see. One useful (if slightly morbid) way to learn more about the adult mayflies and caddisflies emerging from your pond is to check spider webs that overhang the pond each morning. Within these silken traps, you may spot the half-eaten remains of caddisflies and mayflies that emerged from your pond in recent hours.

How will they arrive?

In most cases, the adult life stages of these pond animals will find your pond before or after mating. If your pond is weedy enough and if it is of good enough water quality and size, eggs will be deposited. Often these eggs are laid near the surface of the water and hatch in late summer, thereby overwintering in the pond as nymphs.

CADDISFLIES In all, there are approximately 200 species of caddisfly in the UK and nearly all species feature caterpillar-like nymphs, which make a distinctive 'suit of armour' using a range of items found within the pond (*pictured above*). These items can include bits of grass, sticks, sand, empty snail shells and tiny stones. Occasionally (and very depressingly) they may use bits of discarded plastic. Many caddisfly nymphs stick these items on to their body by first winding a net of sticky thread over themselves.

Some of the most common caddisfly nymphs in garden ponds belong to the genus *Limnephilus*. You can sometimes tell this group of caddisflies apart by the cases that their nymphs build. Some species, such as *Limnephilus rhombicus*, make their cases from specially cut roots arranged at neat right angles. Others, like *Limnephilus nigriceps*, make a Toblerone-like case of three leaves. One particularly common species, *Limnephilus vittatus*, makes its case out of carefully arranged grains of sand. Another common pond species is *Limnephilus flavicornis*. The nymphs of this caddisfly species make a chunky case out of a rag-tag group of pond objects that sometimes includes discarded ramshorn snail shells.

Another caddisfly nymph to look out for in ponds is *Trianodes bicolor*. This is one of the UK's only free-swimming caddisfly nymphs. It uses its long legs to gently row its stick-covered body through the water.

MAYFLIES Mayflies differ from all other insects in that their winged adult stage undergoes a second moult to reach a fully sexual life stage that includes showy body streamers and shining

ABOVE The larval cases of caddisflies can include seeds, sticks and (sadly) bits of plastic.

OPPOSITE TOP A perching adult caddisfly. Note the slightly hairy wings.

wings. Of 50 or so species in the UK, 10 mayfly species regularly use ponds as a nursery habitat for their nymphs. The most likely mayfly likely to visit your pond is the pond olive mayfly (*Cloeon dipterum*), whose nymphs measure about 9mm and possess three 'tails' each with black banding along the middle. Research by the Freshwater Habitats Trust suggests that this mayfly may be found in as many as half of all ponds. Unlike many mayflies, this species lives for up to two weeks as an adult.

ALDERFLIES Some ponds can be home to nymphs of the mud alderfly (*Sialis lutaria*), which fall from eggs laid on the leaves of overhanging vegetation. These nymphs possess three pairs of legs, behind which are rows of feathery gills. They can live in ponds for up to two years, feeding on small invertebrates.

Moths

There are also moths whose larvae specialise on pond plants. These include caterpillars of the brown china-mark moth (*Elophila nympheata*), which construct a protective jacket made of pondweed leaves held together with silk. As newly hatched caterpillars, this species mines the insides of pondweeds and water lilies, along with other moths including *Nymphula stagnata*. Pond dipping during springtime sometimes uncovers caterpillars of *Cataclysta lemnata* – this species often makes cases of duckweed leaves, which it holds together with silk.

FROGS AND TOADS

Garden ponds have become a vital lifeline for many of the UK's widespread amphibians, particularly frogs. These human-made wetland habitats have helped dampen national declines in these species caused by a loss of ponds across the wider countryside. Theirs is a conservation win worth celebrating.

Of four native frog and toad species, only two species are likely to visit your pond. These are the common frog (*Rana temporaria*), a species that does very well in all sorts of ponds, and the common toad (*Bufo bufo*), which tends to prefer larger ponds only. Though these widespread frogs and toads are highly variable in their markings and colours, you can easily tell them apart by looking for a few simple differences.

• **Common frog:** has stripy legs, smooth skin and a clearly visible black marking behind the eyes that looks a little bit like a mask. Also, its eyes have a round pupil.
• **Common toad:** is warty and has a pair of lumps behind the eyes, which are its poison glands. The common toad's eyes are a gorgeous shiny bronze colour and within each eye sits a horizontal, rather than circular, pupil.

It is sometimes easier to tell frogs and toads apart by looking at their behaviour when startled. When disturbed, many frogs leap energetically away to find safety. Toads, however, often just sit still, eyeing up the threat with an almost resolute and grumpy demeanour.

HOW DO FROGS AND TOADS FIND PONDS?

The smell of ponds is very important to both frogs and toads. In spring, both species are drawn towards the smell of ripening algae – an important food source for their tadpoles in the early days of their development. Once they near the pond, other senses kick in. Frogs and toads use their eyes to assess landmarks and obstacles, and their simple ears to listen for the sounds of males calling. Male frogs make a groaning noise in spring, whereas male toads make a 'yip' noise that sounds a little like a squeaky dog toy.

Common frogs are far better at colonising new ponds than toads, the majority of which will return to the same pond each year. Frogs may move many hundreds of metres during their migration back to ponds in spring, but toads are far more enduring. Over a period of some nights, common toads may travel more than 1,000m from overwintering sites back to their breeding ponds.

BELOW A common frog with distinctive dark grey or black area behind the eyes.

ABOVE A female common toad with male attached in a mating position called amplexus.

EGGS AND TADPOLES

During the breeding season, nothing rivals the soap opera that amphibians bring to ponds. So all-consumed by their sex lives do frogs and toads become that it is easy to observe them from very close up at this time of year, allowing you to take in one of the UK's most easily accessible and memorable wildlife experiences.

It is easy to tell frogs and toads apart just by looking at their spawn. Common frogs produce the classic blobs of spawn that many people remember from their childhood. Common toads, however, lay their eggs in long strings, which they wrap around pond plants.

Though superficially similar, frog tadpoles are brown and, upon closer inspection, spotty, whereas toad tadpoles are jet-black. Interestingly, their tadpoles behave differently too. Common toad

tadpoles occasionally form a shoal, rather like fish.

Depending on the size of your pond, how shady it is, and how much food there is available, you can expect tadpoles of both species to develop fully and metamorphose into tiny froglets and toadlets within 10–16 weeks of being laid as eggs in the pond. In particularly dry years, froglets and toadlets may stay at the damp pond edge for days or weeks, awaiting heavy spells of rain that allow them to radiate out into the terrestrial landscape to find feeding opportunities elsewhere.

UPS AND DOWNS

Because wildlife gardens can be such an important habitat for invertebrates, some gardens can support quite large populations of hungry frogs, sometimes 30 or more. In springtime this can sometimes see ponds quite literally become filled with spawn. This is a cause for concern for some people, who worry that their pond will be 'overrun' with frogs in future years. Fear not: amphibians all over the world regularly undergo boom-and-bust cycles. In boom years, diseases may limit frog numbers (see page 125) or food availability. In fact, many frogs may suffer signs of malnutrition and fall foul of hungry predators when their populations skyrocket. Tadpoles, too, suffer from overcrowding and competition for resources. Many tadpoles will eat one another in such situations. All of these factors combine to cause populations to plummet back down after boom years.

NEWTS

Newts are secretive predators that often go unnoticed by pond owners. The UK is home to three species, each one capable of visiting (and sometimes becoming resident in) garden ponds.

By far the most frequently encountered newt is the smooth newt (*Lissotriton vulgaris*). Typically around 110mm in length, this species appears relatively unfussy about the ponds that it visits. Many smooth newts take up residence in wildlife ponds for a few weeks or months, making the most of the glut of tadpoles or small pond invertebrates. Other smooth newts move onto land in the summer months where they seek out logs, stones or patio slabs under which their terrestrial invertebrate prey can be found. In this terrestrial phase, smooth newts have a velvet-like texture to their skin.

ABOVE/OPPOSITE **1** A smooth newt. **2** Palmate newt (male) with 'boxing glove' hind feet and long pointy tail filament. **3** Male great crested newt with spiky body crest and a slight dash of white along the tail.

The palmate newt (*Lissotriton helveticus*) looks very similar to the smooth newt but, at just 90mm, is slightly smaller in size. In the breeding season, males of this species differ from the smooth newt in possessing an unusual tail that has a long filament running out of the tip. The palmate newt also possesses dark webbing on the toes of its hind feet. This newt species favours heathland, grassland and moorland ponds. It is therefore much less likely to be found in most garden ponds.

The rarest and most threatened is the great crested newt (*Triturus cristatus*, the largest and most startling British newt. This dinosaur-esque amphibian grows to 160mm in length and is covered in lots of tiny warts, particularly along its flanks. During the breeding season, males develop a jagged crest and a distinct white 'flash' along the tail.

The great crested newt is the only newt whose habitats are protected by law. If you think you have this newt in your pond, inform your national statutory conservation agency (page 155), which will inform you of what to do next.

UNCOVERING THEIR SECRET LIVES

The activities of newts are best observed after dark. By scanning a torch over the pond, you may see them moving through the water like tiny crocodiles as they search for invertebrate (and tadpole) prey. During the breeding season it is easy to lose hours engaged in their complex interactions, which involve males displaying to females via pheromones sent through the water with an agitated wag of the tail. In the weeks that follow, you may see the egg-laying behaviour of females, which takes place on successive nights for weeks on end. The female great crested newt may lay up to 300 eggs in a single season, each of which she wraps up individually in leaves and other pond detritus. Incredibly, a female smooth newt can lay double this number.

Larvae of all newt species resemble the adults in basic form but, of course, they are dramatically smaller and look more transparent and fish-like. Great crested newt larvae have distinctive black spots, particularly along the tail.

HOW DO NEWTS FIND PONDS?

Newt larvae (particularly those of the great crested newt) are often preyed upon by fish, so newts may actively seek out ponds without fish for their egg-laying. For this reason, garden wildlife ponds are in many ways perfect for newts. All newt species are likely to use smell to find suitable ponds although, interestingly, the smooth newt is also known to be responsive to the sound of calling common toads. It may be that smooth newts use this audio cue as a way to home in on a suitable pond during spring migration.

Amphibian diseases

Amphibians can suffer from a range of diseases, two of which are currently giving conservationists great cause for concern. The first is the fungal disease chytridiomycosis, which is thought to be the smoking gun in a number of amphibian extinctions around the world. The second disease causing concern is a frog virus (a ranavirus) that originated in North America, which causes sudden 'explosive die-offs' in common frog populations in mainland Britain. When afflicted by this disease, many frogs turn up dead in only a matter of days. Often this occurs in the summer months. These diseases are one reason why many wildlife experts advise that frogspawn should no longer be moved from pond to pond. Thankfully, neither of these diseases are a threat to humans or pets.

3

SNAKES AND LIZARDS

The UK is home to six reptile species, all of which can be occasional visitors to wetlands. These reptiles occasionally bask on the banks of rivers and ponds, warming themselves up for a day's hunting. However, only one reptile species – the grass snake (*Natrix helvetica helvetica*) – will regularly move into the water to search for its primary prey, amphibians. This is the reptile species most likely to visit your wildlife pond. If you see one of these impressive snakes moving through your garden towards your wildlife pond, take a moment to enjoy the spectacle. For many people, sightings like these might be the only wild snake they get to see in their whole life.

GRASS SNAKE In almost every case, a snake spotted in or near water in Britain will be the grass snake, our only wetland-adapted reptile. Grass snakes have a bright yellow collar behind the jaws and they can reach an impressive length of 1.5m. However, most of the grass snakes you may come across will be far smaller. Garden ponds are occasional stop-off points for hungry juvenile and subadult grass snakes. This reptile occurs patchily across much of England, Wales and, more recently, southern Scotland.

The grass snake is a non-venomous snake that hunts amphibians, fish and, occasionally, fledgling birds. It poses no threat to people or their pets. When disturbed, the grass snake's panic response is to race away to dense undergrowth. If cornered or lacking an escape route, some grass snakes play dead, particularly if they have been roughed up by pets beforehand. Often, they discharge a smelly paste-like substance by way of a last hurrah.

Grass snakes hibernate underground and emerge from their slumber on warm days in late spring, usually in April. They spend a great deal of their time basking in these early months (*pictured below*), gearing themselves up for breeding, which takes place in June and July. Female grass snakes lay between 8 and 40 eggs in warm piles of rotting vegetation. In late August and September, miniature versions of the adults hatch from these eggs. Adding a compost heap near to your pond is an excellent way of

attracting nearby grass snakes. You may see their tiny offspring (just 15cm or so long) hunting froglets and juvenile newts later in the year.

The UK is home to two other native snake species, neither of which is likely to be seen in your garden. The first is the adder (*Vipera berus*), which has a distinctive zigzag marking running along its back. The second is the highly localised and incredibly rare smooth snake (Coronella austriaca). Both of these snakes are not normally found in water.

LIZARDS Other reptiles often thrive in areas with lots of different microhabitats that provide cover from predators and plenty of prey by way of invertebrates. For this reason, gardens are becoming increasingly important habitats for two widespread reptile species – the common lizard (*Zootoca vivipara*) and the slow-worm (*Anguis fragilis*), a legless lizard.

One under-appreciated wildlife resource is a rockery near your pond, in which there are lots of gaps where reptiles can hide. But there are other garden features that reptiles, particularly lizards, like. Compost heaps are particularly valuable to

ABOVE **1** A male adder with distinctive silver / black markings. **2** Common lizard. **3** Slow-worm showing off its paddle-like forked tongue.

How will they arrive?

Because reptiles are wary of roads and other human infrastructure, most gardens are unlikely to experience visits. However, wildlife corridors are very effective for these species. Hedgerows, railway embankments and allotments frequently have populations of at least some species, a tiny proportion of which may discover your nearby wildlife garden and choose to stay put for hours, days or (in the case of slow-worms) sometimes years. You can help wildlife move through urban areas by encouraging neighbours to add gaps (often called hedgehog holes) to their garden fencing.

slow-worms, for instance. These legless lizards move through them in search of slugs, ants and woodlice. Dotting a few old pieces of carpet around your pond can provide good places for reptiles to bask in early spring. Many will hide underneath them later in the day to hoover up any invertebrates seeking shelter.

BIRDS

Your wildlife pond will be a life-giving resource for local birds, and many species will come back to it again and again for drinking, bathing and feeding. In fact, over time you may notice the same individuals returning each day, giving you an insight into the daily lives of these wonderful animals in your local patch. However, the regularity with which birds visit ponds does change throughout the year according to the weather and the breeding activities of each species. Long hot summers may see ponds visited with desperate regularity.

In winter, birds continue to need water regularly. Keeping a small part of the water ice-free during cold snaps can be an act of kindness for local birds. In fact, an ice-free pond during a snowy winter may attract some stunning winter migrants, including redwing (*Turdus iliacus*), siskin (*Carduelis spinus*) and fieldfare (*Turdus pilaris*).

HOW TO ATTRACT BIRDS TO YOUR POND
Having feeders near your pond is a wonderful way to attract local birds, but be sure to locate them at least a few metres from the pond so that spilled food (and droppings) don't end up fouling the water. Nest boxes located nearby are another recommended option.

POND SPECIALISTS Some larger ponds may become feeding stops for a number of specialist wetland birds. The grey heron (*Ardea cinerea*) is well known for its habit of emptying ponds of their ornamental fish, for instance, but this elegant predator is also partial to frogs, many of which favour garden ponds. The bright blue flash that signals a passing kingfisher (*Alcedo atthis*) may be spotted in some garden ponds, particularly those near canals and rivers. And then there are the specialist birds that favour dense and expansive patches of emergent wetland vegetation – the sedge warbler (*Acrocephalus schoenobaenus*), reed bunting (*Emberiza schoeniclus*) and reed warbler (*Acrocephalus scirpaceus*).

The largest ponds may become attractive spots for ducks, geese and swans. Large ponds with patches of dense vegetation can also become attractive to moorhen, coot (*Fulica atra*), little grebe (*Tachybaptus ruficollis*) and great crested grebe (*Podiceps cristatus*).

LEFT Birdfeeders placed close to your pond are a great way to encourage visitors – but watch out for excess seed waste finding its way into the pond.

The **pied wagtail** (*Motacilla alba*) is a common and widespread bird that often visits ponds to feed on insects, including mosquitoes and midges, as they emerge from the pond surface.

The **robin** (*Erithacus rubecula*) is a year-round resident that often undertakes repeat visits to garden ponds throughout the seasons. It may use nearby trees as a perch from which to sing.

Sadly, the **house sparrow** (*Passer domesticus*) is no longer as common as it once was. Local flocks may boldly sweep into your garden from time to time for a wash and a drink.

When froglets and toadlets emerge from ponds en masse, they often fall prey to hungry birds. These include the **blackbird** (*Turdus merula*), a predator known to make repeat visits to the pond side to seize unwary prey.

Large birds like the **woodpigeon** (*Columba palumbus*) often visit ponds to drink, particularly in the summer.

If your garden has apples, pears or cherry trees, you may also receive a visit from the **bullfinch** (*Pyrrhula pyrrhula*), another year-round resident.

MAMMALS

In the dead of night, a curious cast of mammals may visit your pond. You might need a special hide or a camera trap to spot them or, at the very least, a warm cup of tea and a healthy dose of patience.

HEDGEHOG Sadly, the hedgehog (*Erinaceus europaeus, photo 1 above*) is a species whose fortunes are dwindling, particularly in the wider countryside. Currently, their population stands at just 3 per cent of what it was in the 1950s.

Gardens and their wildlife ponds can help hedgehogs. Garden ponds provide water and food by way of froglets and toadlets, just one of many food items that hedgehogs eat. Each night a hedgehog might eat 70g of prey. Next morning, their characteristic sausage-like droppings litter the edges of some ponds.

Hedgehogs are very capable swimmers but ponds with steep edges can prove very difficult for them to escape. If your pond has unfriendly edges for wildlife, make a makeshift ramp out of stones or some little steps out of the pond using carefully arranged bricks.

WATER SHREW The water shrew (*Neomys fodiens*) is the UK's largest shrew. It spends most of its life living in wetlands, occasionally diving into ponds while searching for nymphs of caddisflies and mayflies to eat. Some take on amphibians many times their size. Water shrews live in small burrows dug into the banks of lakes, rivers and ponds.

FOX In both urban and countryside settings, the red fox (*Vulpes vulpes, photo 2 above*) often visits ponds looking for a quick drink, particularly at dusk or at dawn. In spring, it's been said that some red foxes have been known to claw out frogspawn, which they then devour at the pond edge. They also catch and regularly prey upon frogs.

BATS Bats regularly hunt over ponds, picking off newly emerging mosquitoes and midges and sometimes larger prey, such as caddisflies. The bat you are most likely to see above your pond will be the common pipistrelle (*Pipistrellus pipistrellus, photo 3 opposite*), a small bat that flutters around rather untidily and can be easily spotted at dusk during summer and autumn.

Larger ponds may become frequent hunting sites for Daubenton's bat (*Myotis daubentonii*). This medium-sized bat hunts along the surface of the water, consuming emerging invertebrates within seconds of their metamorphosis. It can even land in the water, swimming for short periods before taking off again upon broad, powerful wings.

OTTER Unless your garden is large and situated near a river, the otter (*Lutra lutra*) is unlikely to visit your pond. However, it may be that the actions of wildlife gardeners one day come to assist the otter in its recovery after years of persecution. Although 90 per cent of the otter's diet is fish, they regularly eat pond crustaceans, molluscs and amphibians, including frogs and toads. In fact, otters are well known for their ingenious methods for getting around the toad's toxic defence. Sometimes they peel off toad skin like a banana before eating the nutritious internal organs.

MINK Sadly, in recent years many ponds, rivers and lakes have become home to the American mink (*Neovison vison*). This invasive non-native predator looks superficially like an otter but it has a pointed face and a bushier tail. The American mink swims in a jerky manner compared with the more elegant and seal-like otter. This species is behind a number of native wetland species declines, including the water vole.

WATER VOLE The water vole (*Arvicola terrestris, photo 4 opposite*) used to be one of the UK's most common mammals. It once lived in a range of wetland habitats, including on the banks of nearly all of our rivers, streams and canals.

Today, ponds provide a stable habitat for many populations.

Water voles look rather like field voles (*Microtus agrestis*) but for their larger size (120–260mm in length) and their fur-covered tail and rounded nose. Water voles can turn up near ponds, provided that the pond overlaps with other wetland sites. Look out for neat piles of cut grass stems at special feeding stations and their charismatic piles of little droppings, called latrines.

BEAVER Because of their penchant for damming streams and rivers, the extinction of European beavers (*Castor fiber*) in the sixteenth century is likely to have had a dramatic impact on the rate of formation of new pools and ponds in the wider countryside. This lack of new ponds undoubtedly affected the fortunes of many freshwater species. Thankfully, the European beaver is being tentatively reintroduced to some parts of Britain. The Wildlife Trusts are currently involved in a number of reintroduction trials throughout England, Wales and Scotland. This is not an animal you can expect to see in your garden pond any time soon, but its reintroduction to the wider countryside is worthy of celebration nonetheless.

How to attract mammals

Making your garden accessible is the key to attracting mammals. You can do this by adding small holes and gaps underneath fences if you have them. You can attract some species like hedgehogs by putting out special food for them at night. Some gardens have security lights near a pond, which can flash on suddenly. This can often startle visiting mammals, so you may need to temporarily turn these lights off to get a chance to spot your pond's mammalian visitors.

EXPLORING YOUR POND

BY NOW YOUR NEW WILDLIFE POND should be beginning to thrive. On every sunny day and on every windy day, with each visiting bird or bug, new animals and plants will be colonising the unexplored niches available in your pond. Congratulations, you have become a local hero to wildlife!

At this point, many pond owners choose to carry on with their gentle management of ponds, happy in the knowledge that their makeshift wetland is benefiting local wildlife. But my feeling is that ponds are there to be explored by humans, too. By carefully checking the inhabitants of your pond every now and then, one can gather a greater understanding of the pond ecosystem, the pond communities, animal behaviour, plant development, pond energetics, adaptations, evolution – the list is endless. Ponds are places for discovery. Places to learn new things. Your pond can serve as a classroom to visiting friends, family, children or (to the teachers reading this) students. And the data you gather from your ponds can have national significance via the many citizen-science research projects that exist to study and understand the UK's biodiversity.

In this chapter, we outline some of the opportunities for pond owners to plug in to the science of ponds. We also outline ways to learn more about the animals in your pond, using microscopes and cameras. Lastly, we introduce some fantastically accessible wild ponds for you to gain inspiration for your next big pond project.

POND DIPPING

There is no quicker and easier way to explore your pond than to get yourself a net and give it a gentle swipe through the water. In this section, we outline a number of ways you might get the most from your pond-dipping endeavours.

NETS Nets come in a variety of lengths and mesh sizes. For most sorts of pond dipping, a mesh size of 1mm is about right. A mesh like this will capture small worms and even most water fleas, as well as tiny life stages of many other invertebrates. It will also catch the bigger creatures, like backswimmers and saucer bugs. If you are only interested in larger creatures such as amphibians and fish, then a 2mm mesh is preferable. If you don't have a net, simply reach for a sieve – just remember to give it a good wash afterwards!

Some expensive pond-dipping nets have a firm rim around the head of the net, which means you can thrust it into thick patches of pond plants or even the pond substrate. This can be helpful in larger ponds, but remember that heavy-handed pond dipping in this way can be damaging to pond plants and animals. In smaller ponds, vigorously thrusting a heavy net around may even be damaging to the pond liner. In many cases, a simple sieve or a cheap rock-pooling net could do exactly the same job.

TRAYS When it comes to pond dipping, a white sample tray (or even an empty ice cream tub) is almost as important as your net. The white background of your tub or tray will help you see pond animals more easily. Remember to fill up your tray with some water from the pond before you begin dipping. You can purchase pond-dipping trays from many online shops. If possible, go for the most hard-wearing you can; when full of water, cheap trays can bend and crack when being lifted up and carried.

ABOVE By far the safest way to dip at the pond edge is to crouch down low, remembering not to lean too far over the edge.

LEFT Some key ingredients for a good dipping session.

OBSERVATION TOOLS A white medicine spoon is incredibly helpful when pond dipping. You can use the small 2.5ml end of the spoon to scoop up smaller creatures like water fleas and water mites, and the weightier 5ml end to scoop up tadpoles and other larger animals. A magnifying glass or mobile-phone magnifier (now widely available) are both useful tools to have on hand for trickier identifications. You can carefully move some pond animals to a laboratory or a nearby desk in plastic specimen pots. Just remember to put these animals gently back in the same part of the pond from which they were scooped as quickly as possible.

RECORDING TOOLS There are a host of organisations that may be interested in your sample results (see page 142), so keep a pencil and notebook handy and remember to take lots of photos. At the very least you should note the date, time and pond location, and the numbers and types of organisms you find. From data like this, it may be that you can track how quickly animals are colonising your wildlife pond year on year, or you may be able to make a rough-and-ready calculation of the degree to which your pond is polluted.

What makes a good tray?

The temptation when pond dipping is to get as much as you can in your net and dump it all in your pond-dipping tray with an audible 'plop'. This temptation is understandable, since this would technically give you the most animals. But there is a problem with this style of dipping, and it is that your tray becomes so congested with material that the animals simply hide among all the vegetation and you fail to see anything.

A good pond-dipping tray might be one-fifth vegetation and four-fifths open space. This gives plenty of white background upon which animals can be spotted. Gently pulling your pond vegetation apart will nearly always unveil new animals.

Remember that your pond-dipping tray will often contain both predator and prey. Dragonfly nymphs and saucer bugs may make short work of tadpoles in such cramped conditions. Either make sure you change your water within a few minutes or, with your medicine spoon, scoop out the predators and put them in a different tray.

Water-filled trays heat up quickly when exposed to full sun in spring and summer. Many animals (especially sticklebacks and newt larvae) become stressed in these conditions. Keep your trays shaded and remember to change the water as regularly as possible.

SAFE POND DIPPING

In the modern age, undertaking pond-dipping sessions with groups of friends, volunteers or students will require you to think over the risks involved and how you intend to reduce or completely remove the likelihood of accidents. In this section we explore a number of issues you will need to consider.

POND DIPPING AND TODDLERS

The most at-risk group by far when playing near ponds are those between one and two years old. This is because children of this age have high mobility but very poor coordination. They may easily lose balance at the pond edge and slip or fall into the pond. Even shallow ponds can be lethal from a child's perspective – a pond 50cm deep is the equivalent of a pond 1.8m deep from the perspective of an adult. This is why, for children, climbing out of water can be so much more difficult than it would otherwise appear.

The single most important rule to remember is to never let children pond dip or play near ponds unsupervised.

SAFETY BRIEFING
When it comes to managing risks, this might be the single most important part of your pond-dipping session. Before starting your pond dip, you should gather everyone together to outline the risks and the ways to minimise those risks. This is an important time to inform participants about areas that are particularly steep, slippery or off-limits entirely for safety reasons.

Where is it best to dip?

Pond-dipping platforms often give the illusion that the best spots for pond dipping are in open water. Actually, you'll find many of the most unusual species at the pond edges, either directly in submerged pond plants or among the networks of stems and roots. Gently move your net over these areas, being careful not to destroy or damage the pond plants, and prepare to be amazed.

SLIPPERY BANKS When lots of people are pond dipping at once, some pond edges can suffer from wear and tear. Within a quarter of an hour or less, these sections can become incredibly slippery. Also, some raised banks may begin to collapse if many pond dippers stand too close together while dipping with their nets. To manage this risk, you should regularly check over the edges of the pond during pond-dipping sessions. Rope off these sections if you have to or redistribute pond dippers to safer parts of the pond if the banks are becoming too slippery or uncertain.

REMEMBER YOUR CENTRE OF GRAVITY This is another important one. Standing at a pond edge with a heavy net out-stretched can dramatically alter one's centre of gravity, increasing the likelihood of slips and topples into the water. Children, who often lack experience of holding nets, don't always realise this. During the safety briefing, the single biggest piece of advice you can give is to recommend pond dippers crouch down while dipping, and that they never lean over the pond with their nets.

BRING A FIRST AID KIT … AND A TOWEL Inevitably, sometimes people do fall into ponds while pond dipping and, in nearly all cases, the only thing that suffers is their pride. Some young people can be deeply embarrassed if they fall in, so give them a towel, a hot drink if needed and some heartfelt reassurance that it happens to everyone, even the most skilled pond dippers, at some point in their lives. Next stop: a shower or bath.

WATER-BORNE DISEASES The three main diseases associated with freshwaters are leptospirosis (Weil's disease), hepatitis A and tetanus. You can limit the chances of catching these diseases in four simple ways:

1 Carefully wash hands and equipment (including kitchen sieves and medicine spoons if used) after being in or near the pond, particularly before eating.
2 Avoid ingesting pond water.
3 Ensure all cuts and grazes that might come into contact with water are covered with waterproof plasters or dressing beforehand.
4 Ensure tetanus jabs are up to date.

If you are using hand sanitiser gels, be sure not to put your fingers in water-filled pond-dipping trays or handle amphibians or any other animals afterwards. These chemicals may be toxic to animals.

RISK ASSESSMENTS Risk assessments help you to identify risks and put into action the solutions that control them. Though these formal assessments sound daunting for newcomers, they are an incredibly helpful and potentially life-saving exercise. Each pond has its own unique risks, so you will need to write your own risk assessment, but the headings for a classic risk assessment include:

1 a definition of the hazard;
2 the people at risk;
3 the level of risk (low, medium, high);
4 the action required to minimise or control risk;
5 the level of risk after actions have been applied.

SCHOOL PONDS

School ponds are of immense educational value. They allow opportunities for hands-on practical fieldwork and opportunities for students to interact with nature socially, creatively, scientifically and artistically. However, by having a pond, schools have a duty of care to ensure that students, teachers and anyone else who might be on the school premises remain safe at all times. Managing school wildlife ponds involves making detailed risk assessments and safety policies, which are then drafted and overseen by named staff responsible for their enactment. These formal documents will be required to be monitored and reviewed at regular intervals. An emergency action plan will also be required. For further information on school ponds, search the Royal Society for the Prevention of Accidents website, www.rospa.com.

NIGHTTIME VISITORS

The nighttime brings with it a change of cast akin to what you might expect within a tropical rainforest or coral reef. Within minutes of the setting sun, newts may dip in and out of the water, looking for places to lay eggs. Caddisflies flit in and around the surface of the water, seeking opportunities to mate. Noisy water boatmen sing their scratching songs. And that's just the start.

Here are some pieces of useful advice to help you discover more about the nocturnal goings-on in and around your pond.

TORCHING Torching is by far the easiest way to get a feel for what is happening in your pond at night. Using a hand torch and standing at the edge of the pond, slowly scan across the water's surface with the beam and you may catch glimpses of frogs and newts below. You may also spot swimming leeches, numerous

water beetles and surface-dwelling bugs such as water crickets. Professional ecologists and wildlife conservationists buy special high-powered torches for this purpose, which have become quite accessible online. The best specialist torches are in the range of 500,000 to 1 million candlepower. Going brighter than this is unnecessary. It might even be detrimental and disturbing to pond life.

CAMERA TRAPS Camera traps have never been cheaper and more accessible than they are right now. These hard-wearing cameras can be left outside and tied to a wooden stake or tree where they take photos of anything that strays past, from frogs to foxes. Simply download the camera's memory card each day and, hey presto, you have a diary account of the comings and goings of the larger nocturnal animals visiting your pond.

FOOTPRINTS The soft, wet soil around the edges of some ponds can make a useful tracking bed. Before turning in for the night, smooth out the bank's mud with a branched stick or rake so that it is nice and smooth. Next day you might find a variety of tracks, including red fox, hedgehog, frog or, if the ground is soft enough, the faint impression of a passing newt with its tail dangling behind as it lumbers towards the pond.

Be safe!

Do not visit larger garden ponds on your own at night! Go with a friend or family member, ideally wearing hi-vis jackets so that you can find one another easily. Remember the risks that surround ponds and manage them properly.

MOTH TRAPS Like camera traps, moth traps have also become a very accessible bit of kit for amateur naturalists. Most moth traps are capable of temporarily attracting a number of night-flying pond insects, including adult caddisflies, water boatmen and saucer bugs. For safety reasons, remember not to situate your moth trap too close to the pond edge – electricity and water do not mix. Also, refrain from leaving your moth trap on for too long. You won't want to interrupt the nocturnal goings-on within your pond any more than you have to.

THROUGH THE LENS

One of the most startling things about ponds is the strange and complicated hidden workings that go on at a microscopic level. Under a powerful microscope, even the most transparent drop of pond water can reveal itself to be a carnival of wandering, bumbling, spiralling and spinning tiny organisms. In this section, we focus on the microscopes and other magnifying devices that help you shed light on the smaller creatures in your pond.

BINOCULAR MICROSCOPES
On the whole, microscopes come in two forms – binocular and light. Of the two, binocular microscopes are better for looking at many invertebrates because they offer more of a 3D image, which is helpful for when you are trying to manipulate specimens into a good position to be inspected. Large binocular microscopes offer unrivalled opportunities to examine the close-up workings of pond creatures, but they often come with a hefty price tag. However, because they have few moving parts, most binocular microscopes last for generations.

LIGHT MICROSCOPE
Many light microscopes offer you views of pond wildlife using a far greater magnification. Some light microscopes will allow for 500x magnification, easy enough to see pond life on a cellular level. Nematode worms, rotifers and water-bears (more properly known as tardigrades) are just three types of diverse and frankly mind-blowing animals whose lives you can explore using this form of microscopy.

USB MICROSCOPES
USB microscopes have become a dramatically cheaper alternative for those interested in entry-level pond microscopy. Nowadays, there are a great many options out there and many of these microscopes can offer up to 200x magnification, which is enough to study in pretty impressive detail the basic anatomy of smaller creatures such as water mites and water fleas, including *Daphnia*.

are simple to use and remain relatively cheap. Some even plug into mobile phones, allowing you to quickly post close-up photos of your pond animals to social media, should you be that way inclined. Check online reviews before purchasing your borescope to double-check that it really is waterproof.

LOUPES You probably won't have a plug socket available at your pond, so you may have to rely on more traditional forms of magnification for close-up observations of pond animals in situ. A must-have bit of kit is a loupe. A loupe is a simple lens that folds in and out of a protective case. Some loupes offer a range of magnifications, with lenses of different powers folding up alongside one another like the tools in a Swiss Army knife. Loupe magnifications vary, often from 5x to 20x. For most kinds of animal investigations, a magnification of 5–10x is more than adequate.

LIGHTING Unruly surface reflections can cause microscopes all sorts of problems with glare and focus, so lighting is really important when investigating smaller pond animals and plants. This is especially true for those using cheaper USB microscopes.

Many microscopes offer backlights or mounted lights to counter this problem, but you could also consider purchasing an extra 'bendy' light (many USB ones exist) to light your subjects in the best way possible. Some USB microscopes have a special polariser dial to help alleviate this problem. These are incredibly effective but they come with a price.

Remember that many lights give off heat. If you are studying live objects, minimise the disturbance this causes them by operating quickly and putting the animals back in the pond as soon as possible.

BORESCOPES Borescopes are essentially a camera on a stick that can be poked into hard-to-reach places, like within the labyrinth-like roots of pond plants, in rockeries or underneath large rocks within the pond. Many borescopes

WATCHMAKER'S LENS Some pond dippers prefer to use a classic watchmaker's lens to see things up close. With a bit of practice, you can grip this lens in place by using your cheek and eye muscles. This allows you to retain the use of both hands, provided you maintain a pirate-like scowl throughout.

PLUGGING INTO THE SCIENCE

Being accessible yet self-contained habitats, ponds are ripe for acquiring very useful scientific data. Comparisons between lots of ponds allow scientists to spot trends over time in communities, populations, pollution and behaviour, as well as much else. The observations from your pond have the potential to plug into national initiatives that help scientists learn more about these precious wetland habitats. Here's how you can contribute.

RECORDING SPECIES DATA One of the best things about pond dipping is the thrill of not knowing what you might find with each sweep of the net. This 'lucky dip' aspect to pond dipping keeps it exciting for all involved. But there is more fun to be had, for the data you collect while dipping also has immense value to science.

The details of the creatures present in your pond (and the creatures that are not) can provide important information about national distributions of species and, for instance, how they may be moving northwards through climate change or declining year by year through the continuing fragmentation of natural habitats due to human influence. When recorded, these species observations are often referred to in plain English as 'records'. Records provide the baseline information about wild animal distributions.

Generally speaking, a species record normally consists of the four 'w's:

- **What** did you see? (species name/possible species ID/photo)
- **Where** did you see it? (grid reference/pin on an online map)
- **When** did you see it? (date and time)
- **Who** saw it? (your name, contact details)

You can give this information to Local Records Centres, of which there are many throughout the UK. Your local Wildlife Trust is likely to be working very closely with your Local Record Centre, so the Wildlife Trust should be your first port of call if you can't easily find their details. Websites like iSpot (www.ispotnature.org) allow

you to provide species records quickly and easily while in the field. Within days, provided you have photos, many of your observations will be double-checked and formally identified by experts in the field.

An additional option is to contribute to a recording scheme that specifically focuses on pond species. The Freshwater Habitats Trust has a project called Pond Net, which sees networks of pond spotters recording and monitoring the fortunes of widespread, often common species in ponds. Get involved through the Freshwater Habitats Trust's website (see page 154).

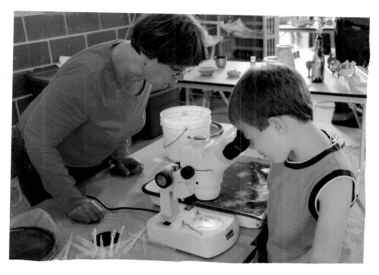

SEASONAL DATA Dates of key events in your pond can provide very useful data for those scientists studying phenology – the timing of cyclic and seasonal natural phenomena. By accruing data from lots of ponds across the UK, these scientists can plot the timings of natural events year on year, helping assess the rate at which the British climate is changing. Phenological research is becoming an important tool for gauging how well species may or may not adapt in the face of climate change, so your records really do count. One easy way to contribute phenological data from your pond is to record the first occurrence of frogspawn. You can report this (and other phenological data) via the Nature's Calendar website, which is managed by the Woodland Trust (see naturescalendar. woodlandtrust.org.uk).

JOIN A LOCAL VOLUNTEER GROUP

One fun and sociable way to secure a healthy future for freshwaters is to join a local group of conservation volunteers. Many of these local groups will have opportunities to contribute to pond-digging or pond-renovation projects, as well as offering often lively and wide-ranging talks about local nature in the winter months. Your local Wildlife Trust will have details of local opportunities. These groups can be especially useful to younger people looking for opportunities to find their way into the wildlife-conservation sector.

If amphibians are more your thing, you could also search out your local Amphibian and Reptile Group (ARG). Like Wildlife Trust local groups, these independent volunteers offer opportunities to plug in to local practical pond projects, such as digging new ponds and surveying threatened species like the great crested newt. These groups can also offer training. Many ARGs work very closely with, or are even coordinated by, Wildlife Trusts.

ENHANCE YOUR KNOWLEDGE – TRAINING

Your local Wildlife Trust offers a range of training opportunities run by local (and often national) experts in a variety of fields. These training sessions allow you to gain a practical understanding of oft-overlooked groups of species, as well as providing crucial identification tips. Species identification courses may include water beetles, water bugs, dragonflies and damselflies, and other fascinating groups of pond animals.

TEMPORARY HOUSING

Temporarily keeping some pond animals in a tank is not generally advised but there may be times when it becomes necessary. This may be, for instance, while you are cleaning up a small garden pond and you have sensitive animals like fish or tadpoles that don't like being in buckets for long periods. Alternatively, a temporary tank may be necessary for photographing or identifying certain species, or for educational purposes.

CHOOSE A TANK Your first option is to consider your tank. Plastic tanks are easier to carry and maintain but regular cleaning can lead to scarring on the inside surface, which can make getting a clear shot with a camera that bit more difficult. Plus, if a plastic tank cracks, it's done for because gluing is not really an option. Glass tanks are hard-wearing and, after cleaning, they remain crystal-clear. However, glass tanks cost

more and they can be very heavy. For all tanks, generally speaking, the bigger the better. Fairly shallow and wide tanks are better than tall, thin tanks. This is because oxygen diffuses more readily into tanks with a good surface area per unit volume.

Upon purchase, you will need to give your tank a good clean to remove any chemical products that may have been applied during its creation. Use an old toothbrush for awkward corners and rinse it out many times before use.

ADD GRAVEL Pea gravel is the type of gravel most commonly used in aquaria. It is easily available from many garden and aquatic centres, though you will need to wash your gravel out in a bucket before first use. Keep swilling the gravel in this water, grinding the pieces against one another until the water turns brown. Remove the water and refill the bucket. Swill the gravel again until the water turns brown. Repeat this process until the water remains clear. Only at this point is the gravel ready for your tank.

PLACE YOUR TANK

If you're not intending to use an artificial light source, put your tank somewhere where it won't be in direct sunlight for long. Water in tanks can heat up very quickly, which can stress some pond organisms. You will need to rest your tank on an even surface. When full, it will become very heavy and the bottom side may crack if it isn't supported across its entire surface.

FILL UP YOUR TANK

Always use water from your pond to fill up your tank. This is the water that your pond animals have adapted to, so it will minimise stress to the animals involved. Gently pour this water onto a (washed) plastic bag, which you have placed over your gravel. This will reduce the chances of washing any residual dirt and dust into the tank, which will make the water cloudy. You may want to place a net or cover over the top of the tank to stop beetles and other flying water insects escaping.

POND PLANTS

Take some of the submerged water plants from the pond and arrange them in your tank. Make at least one area quite dense, so that pond animals have plenty of places to hide if spooked. These plants will add some much-needed oxygen to the water.

REMEMBER THOSE FOOD CHAINS!

In the wild, predators and prey regularly interact but in your temporary tank, the animals become your responsibility. At all costs, avoid putting predatory insects or fish in with smaller pond animals, particularly tadpoles. If need be, create two separate tanks: one for predators, one for prey.

Some top tips:

• You won't need a pump or filter for your temporary tank if you follow the steps above.

• Don't allow inquisitive pets near your tank. They may choose to drink out of the tank or, worse, in the case of cats, intentionally fish out your pond animals to eat.

• To avoid accidentally stressing animals, do not keep pond animals in your temporary tank for more than 24 hours. If it is a hot summer's day, anything more than 12 hours is not advised.

YOUR POND AND ITS FOOD WEBS

Ponds are habitats from which many famous ecologists first gained inspiration and wonder for the natural world. By watching the goings-on of pond inhabitants, one quickly grasps how energy moves through ponds via a web of interacting organisms, arranged in tiers from producer to predator – the so-called food chain. But food chains are about more than just academic study. Understanding your pond's food chains can give you the diagnostic tools to solve problems your pond may come to have in the future, such as apparently endless blooms of freshwater snails or a dwindling number of tadpoles year on year.

You will likely remember food chains from school. They are chains of organisms dependent on one another to power their energetic needs. At its simplest, a food chain in your pond may be, for instance, algae (a primary producer) being eaten by tadpoles (a primary consumer), which are preyed upon by smooth newts (the secondary consumer). Food chains are a helpful way to picture the relationships between the animals in your pond but, in reality, things are far more complex. For instance, algae are hoovered up by many other organisms, such as water fleas, flatworms and snails, with whom tadpoles must compete. And tadpoles don't only depend on algae. They also consume dead animals, particularly in the later stages of their development. Plus, they occasionally feed upon one another. Lastly, newts aren't always 'top dog', so to speak. Adult newts, as well as fish and dragonfly nymphs, all consume newts' earlier life stages.

When viewed in more detail like this, ponds become a mess of interacting organisms all with their own unique and ever-changing energetic requirements. But you can make sense of this mess by scientific observation. Closer inspection of pond relationships exposes a 'web' of interacting animals rather than a simplistic chain.

THE BOTTOM OF THE CHAIN
A single drop of pond water put under a microscope exposes the multitude of incredible forms that carry the status of primary producer. Many of the most beautiful are micro-algae known as diatoms. Together with diatoms in the oceans and soils, diatoms in freshwaters generate about 20 per cent of all oxygen produced on the planet each year. They are incredibly important to the workings of many of Earth's ecosystems. And they are not alone, for there are other species of algae that play an equally important role as primary producers in ponds.

MID-LEVEL INTERACTIONS On land, vertebrates such as amphibians, birds, reptiles and mammals feed upon most insects and spiders. In ponds, however, the tables are somewhat reversed. A range of predatory invertebrates hunt with stealth or actively chase young vertebrates, particularly the tadpoles of amphibians and the larvae of fish and newts. Among the most nightmarish of these invertebrates are great diving beetles (and their larvae), dragonfly nymphs, water scorpions, water bugs and the water spider. All of these invertebrates are able predators, capable of dispatching vertebrate prey with their own unique brand of cold-blooded cool.

DETRITIVORES Each year, dead animals and plants in your pond (including fallen leaves from trees and shrubs) will provide food for something else – the detritivores. Mostly, the detritovores in your pond will be bacteria or fungi, but there are detritivore animals that are likely to undertake this role in even the smallest of ponds. These include water hog-lice, water shrimps and flatworms.

PARASITES Though the word alone may send a shiver down the spine of some readers, parasites make up a great deal of the biomass in freshwater ecosystems. In estuaries, for instance, the biomass of flukes (parasitic trematode worms) may actually exceed the biomass of the estuarine birds that reside there. In your pond, most of the tiny parasites will be hidden from view, deep within the bodies of the animals (particularly fish) they parasitise.

Dragonfly

Iris

Gnat

Common pond skater

Whirligig beetles

Froglet

Common frog

Water scorpion

Water stick insect

Spiked water milfoil

Lesser water boatman

Leech

Water fleas

Great ramshorn snail

Caddisfly larvae

Great pond snail

Newt

Great diving beetle

Dragonfly nymph

GLOSSARY

aestivation period of dormancy undertaken by some animals during hot or dry spells of weather, often in summer

biodegradable something capable of being decomposed by bacteria or other organisms

carnivore animal that preys upon or feeds upon the flesh of other animals

chytridiomycosis disease known to amphibians, which is caused by infection with the chytrid fungus *Batrachochytrium dendrobatidis*

crustacean member of the Order Crustacea. In ponds, these are mostly hard-shelled aquatic creatures including freshwater shrimps, water hog lice, seed shrimps and water fleas

decomposition breaking down of organic matter into smaller and simpler substances

desiccation process of extreme drying

detritus materials that have been eroded and washed away in smaller pieces, often referring to gravel, sand and silt. Can also refer to decomposing/decomposed organic matter.

ecosystem biological community made up of interacting organisms that share their physical environment

erosion process through which rocks and soils are gradually worn down or washed away

gland organ or cell within an animal body, capable of secreting specialised chemical substances often used for survival or body maintenance

habitat environment in which an organism (or collection of organisms) spends at least part of its life

herbivore animal that feeds upon plant matter

herpetologist scientist who studies amphibians and/or reptiles

hibernacula place in which an animal seeks refuge, often seasonally. Commonly, this term is used when referring to the overwintering locations of amphibians and reptiles.

impermeable material or membrane through which liquids or gases cannot pass

invasive term used to describe non-native species capable of spreading and outcompeting native species

invertebrate animal that lacks internal bones and has no spinal cord

larva (plural: larvae) distinct juvenile life form that often precedes the adult life stage. Most larvae occupy a different habitat or undertake a different way of life to the adult life stage.

megafauna word commonly used to describe large animals often more than 1 tonne in adult weight

metamorphosis distinct transformation stage that many invertebrates go through between juvenile and adult life stages

mulch mixture of decomposing leaves and straw that can be used intentionally to enrich soils

native organism found naturally or specifically within a country or area

niche descriptive word used to specify the (often unique) way in which an organism fits into its local ecological surroundings

nitrate derivative of nitric acid, which can be assimilated by plants and algae quickly, allowing them rapid growth

non-native animal or plant not known to be native to a given locality

nutrient any substance capable of providing nourishing qualities to an organism

nymph immature life stage of insects that undergo gradual metamorphosis towards an adult life form. Nymphs are particularly known from dragonfly, damselfly, caddisfly and mayfly groups.

omnivore animal that feeds on more than one obvious food source, for instance, by eating both animals and plants

oxygenate to supply, treat or enrich with oxygen

parasite organism that lives in or upon the body of another organism, drawing nutrients to the detriment of its host

pH measure of the acidity or alkalinity of a given substance

propagate taking cuttings of roots, shoots or layers with the intention to deliberately cultivate the plants elsewhere

puddling to squash freshly laid clay across the base of an unfilled pond, effectively sealing and securing the gaps to minimise leaks

pupa (plural: pupae) developmental stage that occurs between the larval and adult life stages of many insects. This is the life stage in which metamorphosis takes place.

ranavirus genus of viruses (family *Iridoviridae*) known to infect amphibians and reptiles

rhizome continuously growing stem that grows horizontally under the soil, sprouting lateral shoots as it does

silt very fine grains of sand, clay or other soil components deposited on the bottom of a pond, river, stream or lake

stridulate to make a sound by rubbing one body part against another body part. The term commonly refers to insects such as grasshoppers but can equally be applied to lesser water boatmen, males of which rub their genitals against a special patch upon their abdomen.

subsoil layer of soil that lies underneath the surface layer

topsoil topmost layer of soil, often rich in nutrients

toxin any chemical that is poisonous to another organism. Often refers to poisons specifically derived from microorganisms that are capable of causing disease.

transpiration loss of plant moisture by evaporation, especially through the leaves

vertebrate animal with a spinal cord. Includes all animals with an internal backbone.

USEFUL ORGANISATIONS

ORGANISATIONS THAT CAN PROVIDE SPECIALIST ADVICE

The UK is home to an enthusiastic and knowledgeable collection of wildlife charities able to offer further help and advice about wildlife ponds, either via their websites or over the phone if necessary.

AMPHIBIAN AND REPTILE CONSERVATION TRUST (ARC)

The ARC is a wildlife charity that represents the interests of the UK's amphibians and reptiles. As well as managing a suite of important sites for rare species, the ARC is also a major player in reintroducing once-lost native species, like the pool frog, back into the wild. **www.arc-trust.org**

AMPHIBIAN AND REPTILE GROUPS OF THE UK (ARG UK)

These important volunteer groups work across the country. Many offer opportunities to engage in training for local monitoring of amphibians and reptiles, as well as seasonal talks and social events. In addition, there is also a national conference each year (the Herp Workers' Meeting) and a website with plenty of details for local contacts and groups. ARG UK offers a one-stop shop for how to get engaged in local conservation of amphibians and reptiles. **www.arguk.org**

BRITISH DRAGONFLY SOCIETY (BDS)

Founded in 1983, the BDS is committed to conserving Britain's dragonflies and damselflies through scientific research and public engagement. Today the society has more than 1,600 members, many of whom are represented by local volunteer groups. You can find out more about these local groups by visiting the BDS website. **www.british-dragonflies.org.uk**

BUGLIFE

Buglife is a national wildlife charity that represents the interests of native invertebrates, including spiders, insects and crustaceans. The charity's extensive website offers lots of advice and information sheets about invertebrates, including some referring to freshwater invertebrates like water beetles. **www.buglife.org.uk**

FRESHWATER HABITATS TRUST

The Freshwater Habitats Trust works to protect freshwater life for everyone to enjoy. Ponds are a big part of the work that the charity oversees. As well as running citizen projects to monitor water pollution and assess the status of widespread species, the Freshwater Habitats Trust has also put together many easy-to-digest best-practice guides that are easily found on its website. Put simply, there is no better source of scientifically sound technical advice about ponds and how to manage them. **www.freshwaterhabitats.org.uk**

FROGLIFE

Froglife is a national wildlife charity that runs numerous projects across England and Scotland, many involving the creation of new ponds and the renovation of existing ones. As well as providing practical opportunities for both young and old to engage with wildlife in London, Peterborough and Glasgow, the charity also provides garden pond advice through an online section of its website entitled 'Just Add Water'. **www.froglife.org**

GARDEN WILDLIFE HEALTH – INSTITUTE OF ZOOLOGY (ZSL)

This long-running project encourages the public to observe incidents of unusual mortality in garden species, including frogs and toads. Since launching more than two decades ago, the project has helped scientists explore and understand a host of complex diseases that exist in the wild including, in the common frog, the ranavirus disease sometimes referred to as 'red-leg'. **www.gardenwildlifehealth.org**

THE RIVERFLY PARTNERSHIP

This partnership of anglers, conservationists, scientists and water-course managers seeks to conserve Britain's riverflies – the collective term for freshwater insects that include mayflies, caddisflies and stoneflies. As well as providing helpful identification advice, the partnership is also involved in both local and national events. **www.riverflies.org**

THE ROYAL SOCIETY FOR THE PROTECTION OF BIRDS (RSPB)

The RSPB is working towards a healthy environment rich in birds and wildlife. As well as managing a suite of top-class nature reserves (within which freshwaters play a big part), the RSPB has also become an important player in influencing the government to protect native species and habitats, including freshwaters. Its website provides an excellent one-stop shop for guidance on making gardens more nature-friendly. **www.rspb.org.uk**

THE WILDLIFE TRUSTS
Local Wildlife Trusts may be your first port of call to learn more about freshwater animals and plants locally. There are 46 local Wildlife Trusts across the whole of the UK, including on the Isle of Man and Alderney. Most Wildlife Trusts offer training courses as well as running family events and local volunteer projects that may include pond building, pond maintenance and pond surveying. **www.wildlifetrusts.org**

STATUTORY CONSERVATION AGENCIES

NATURAL ENGLAND
www.gov.uk/government/organisations/natural-england
Natural England
County Hall
Spetchley Road
Worcester WR5 2NP
Telephone: 0300 060 3900
Email: enquiries@naturalengland.org.uk

SCOTTISH NATURAL HERITAGE
www.nature.scot
Great Glen House
Leachkin Road
Inverness IV3 8NW
Telephone: 01463 725 000
Email: enquiries@nature.scot

NATURAL RESOURCES WALES
www.naturalresources.wales
Natural Resources Wales
c/o Customer Care Centre
Ty Cambria
29 Newport Rd
Cardiff CF24 0TP
Telephone: 0300 065 3000
Email: enquiries@naturalresourceswales.gov.uk

NORTHERN IRELAND ENVIRONMENT AGENCY
www.daera-ni.gov.uk
Klondyke Building
Cromac Avenue
Gasworks Business Park
Lower Ormeau Road
Belfast BT7 2JA
Telephone: 0845 302 0008
Email: nieainfo@daera-ni.gov.uk

TAKING INSPIRATION FROM NATURE

One simple way to gather inspiration for your own pond is to go and explore the UK's splendid array of showpiece natural ponds. Many of these sites offer free guided walks or regularly hold family fun days. Alternatively, you may prefer to visit these sites when it's quieter at dusk or dawn to enjoy their diverse splendour at a more leisurely pace. Search online for the following sites for further details.

Ainsdale Sand Dunes, Lancashire – Natural England This National Nature Reserve (NNR) never fails to deliver by way of ponds and pond life. The shallow dune ponds found here are home to the rare natterjack toad, which can be seen (and heard) in great numbers in spring by attending special guided walks. Many other protected amphibian and reptile species are also found here, including the great crested newt and the sand lizard (*Lacerta agilis*). For this reason, visitors must stick to signed pathways unless they have a special licence.

Brandon Marsh, Warwickshire – Warwickshire Wildlife Trust Situated just 3km south-east of Coventry, this former gravel pit has blossomed into a rich and vibrant local hub for nature. It features wetlands and reedbeds, wet woodlands and grasslands within which are dotted a range of well-tended ponds. There is also a very accessible and family-friendly bird hide next to the visitor centre, which is always very popular.

Camley Street Natural Park, London – London Wildlife Trust Located not far from King's Cross and St Pancras railway stations, this recently refurbished urban nature reserve offers a range of educational activities, many of which involve pond dipping. Well-informed volunteers are often on hand to help identify tricky finds. This is a must for those wishing to enjoy an hour or two before catching a train.

Castle Espie Wetland Centre, Northern Ireland – Wildfowl & Wetlands Trust (WWT) This bustling 60-acre wetland site south of Belfast is a wonderful refuge for migratory wetland birds as well as bats, which can regularly be spotted feeding over the reedbeds at dusk. As well as networks of small lakes and ponds, the wetland centre at Castle Espie also has an attractive saltwater lagoon that is popular with hunting dragonflies and kingfishers.

Crom, County Fermanagh – National Trust This family-friendly site offers much by way of opportunities for freshwater exploration. Among the paths and waterways there are a number of opportunities to spot threatened species, including red squirrels (*Sciurus vulgaris*) and otters, or maybe even the European pine marten (*Martes martes*). You can even hire boats for an up-close freshwater experience.

RIGHT Trainee pond surveyors putting their new-found skills to the test. Many nature reserves offer bespoke training courses.

BELOW Messy but fun! Pond restoration can be great for team-building.

Decoy Heath, Berkshire – Berkshire, Buckinghamshire and Oxfordshire Wildlife Trust

Decoy Heath is one of the best spots for dragonflies and damselflies in England. Many species are found here, including the downy emerald dragonfly (*Cordulia aenea*), the keeled skimmer (*Orthetrum coerulescens*) and the small red damselfly (*Ceriagrion tenellum*), one of the UK's tiniest damselfly species. Though a car park is present, access is limited, so you will need to call Berkshire, Buckinghamshire and Oxfordshire Wildlife Trust (01635 351157) to arrange your visit.

Hampton Nature Reserve, Peterborough – Froglife

This 122-hectare site (formerly known as Orton Pit) was once home to an enormous brick clay works. Over time, the linear channels of clay dug out from the site filled with water, creating a large wetland network of more than 300 ponds. Today, the site is thought to be the most valuable stronghold for the great crested newt in Europe. This is an incredible must-see wetland site, highlighting how important so-called 'brownfield sites' can be for nature. Access to Hampton Nature Reserve is available through a host of guided walks undertaken by the national wildlife charity, Froglife.

Jupiter Urban Wildlife Centre, near Falkirk – Scottish Wildlife Trust

This former wasteland in the heart of industrial Grangemouth is now a restored wetland paradise. As well as offering regular pond-dipping sessions, the Jupiter Urban Wildlife Centre also has a visitor centre providing plenty of hands-on advice for urban wildlife gardening, including how to make and tend to your own garden wildlife pond.

WWT Llanelli Wetlands Centre, Carmarthenshire – Wildfowl & Wetlands Trust

Situated 8km north of Swansea, this popular visitor centre is a rich network of freshwater habitats including ponds, lakes and estuary pools. In winter, approximately 60,000 birds call the estuary home. The site has accessible pathways throughout for those with limited mobility and is open to trained assistance dogs.

Loch of the Lowes, Perthshire – Scottish Wildlife Trust

As well as being home to ospreys (*Pandion haliaetus*) in spring and summer, this 98-hectare site is also famous for its reintroduced population of beavers, a project that is currently going very well. The site is accessible to wheelchair users, and assistance dogs are welcomed. TV cameras in the visitor centre allow visitors to watch secretive wildlife up close without risk of disturbance.

WWT London Wetland Centre – Wildfowl & Wetlands Trust

This network of former Victorian reservoirs was turned into an urban wildlife paradise thanks largely to the efforts of the WWT. Today, a number of impressive and unusual species are fairly easy to spot among the lakes and ponds. These include grass snakes, gadwall (*Mareca strepera*) and the northern shoveler (*Spatula clypeata*). The site is also home to a noisy and rather boisterous non-native amphibian, the marsh frog (*Pelophylax ridibundus*). A number of accessible pathways exist across the site and there are regular guided walks and events at the visitor centre.

Minsmere Nature Reserve, Suffolk – The Royal Society for the Protection of Birds (RSPB)

With more than 5,000 species records on the reserve in recent years, Minsmere is a national stronghold for a number of impressive wetland species, including otters and water voles. The site has a number of accessible paths, and the large and informative visitor centre has excellent facilities. There is a charge for non-RSPB members visiting this site.

New Forest National Park

In all, the New Forest is home to 1,000 wild ponds. This makes it one of most important places for a variety of threatened animal and plant species, many of which specialise in the temporary ponds that abound here. One in three ponds found in the New Forest National Park contains at least one insect species known to be nationally rare. This statistic alone speaks volumes. Guided walks and other on-site events are highly recommended.

Potteric Carr, Yorkshire – Yorkshire Wildlife Trust

This former fenland was once drained for agriculture but has since re-flooded, creating a mosaic of ponds and small lakes. Among the most celebrated of winter visitors here is the bittern (*Botaurus stellaris*). Fully accessible paths lead you around the site, with plenty of opportunities for birdwatching and pond dipping. It is free to Wildlife Trust members.

LEFT Youth conservation volunteers are an increasing part of today's wildlife scene.

OPPOSITE Many nature reserves run holiday activities for kids and families.

Redgrave and Lopham Fen Nature Reserve, Norfolk – Norfolk Wildlife Trust This fenland site has been restored back to its former glory courtesy of careful management by Norfolk Wildlife Trust. As well as rare birds and dragonflies, some ponds are home to the fen raft spider (*Dolomedes plantarius*), a large and impressively stocky spider known from only a handful of sites in the UK.

Saltholme Nature Reserve, Cleveland – RSPB Saltholme mixes impressive wildlife with a superb family-friendly visitor centre. The hides and screens at the reserve allow regular sightings of a range of flagship species feeding within the wetlands, including water voles, marsh harriers and bitterns. Accessible trails are found throughout the reserve and there is a handy play area to tire out young ones before getting back in the car.

WWT Slimbridge – Wildfowl & Wetlands Trust Situated in Gloucestershire, Slimbridge is the headquarters of the Wildfowl & Wetlands Trust. This wetland nature reserve was established in 1946 by the artist and naturalist Sir Peter Scott, and each decade has seen it improve upon its credentials as an immensely enjoyable and educational family attraction.

Welsh Wildlife Centre, Pembrokeshire – The Wildlife Trust of South and West Wales This family-friendly site offers accessible paths through woodlands, reedbeds and past a number of ponds notable for dragonflies and water beetles. The site also offers good views of kingfishers and otters, the latter of which can sometimes be spotted frolicking in the neighbouring River Teifi.

FURTHER READING

Many technical guides to freshwaters exist, including numerous identification guides to assist you in discovering more about the animals within your wildlife pond. Here are a handful of my favourites, nearly all of which have contributed in some way to the preparation of this book.

Britain's Reptiles and Amphibians (WILDGuides), Howard Inns (Princeton University Press, 2011) This book is an excellent illustrative guide for those eager to learn more about species identification, behaviour and, vitally, the successful conservation of British amphibians and reptiles. Aimed at amateurs and professionals alike this colourful guide is a book I come back to again and again.

Chris Packham's Back Garden Nature Reserve, Chris Packham (Bloomsbury Publishing, 2015) Endorsed by The Wildlife Trusts, this book combines numerous facts about freshwater species (such as midges and mosquitoes) that we overlook all too often. Initially published by New Holland in 2003, this book was updated and reprinted in 2015 by Bloomsbury and it is packed with great advice effortlessly delivered courtesy of Packham's trademark enthusiasm.

Freshwater Life of Britain and Northern Europe (Collins Pocket Guide), M. Greenhalgh and D. Ovenden (Collins, 2007) This is arguably the most accessible and informative pocket guide to freshwater plants and animals that exists. The book includes hundreds of well-drawn illustrations and species descriptions of all major pond organisms you might see in your pond, including tiny worms, mites and seed shrimps. Through my frequent thumbing, I have worn through three copies!

Great Crested Newt Conservation Handbook, T.E.S. Langton, C.L. Beckett and J.P. Foster (Froglife, 2001) Though much of the advice in this technical guide is aimed at encouraging the great crested newt, many of the principles explained apply to wildlife ponds in general – even smaller garden ponds. This book is available free through the Froglife website, www.froglife.org.

The Pond Book: A Guide to the Management and Creation of Ponds, P. Williams, J. Biggs, M. Whitfield, A. Thorne, S. Bryant, G. Fox and P. Nicolet (Ponds Conservation Trust, 1999) To put it simply, there is no better technical guide to pond creation and management out there than this. This book is filled with good advice, particularly for larger countryside ponds, much of which is based on years of scientific study through the fleet of excellent scientists involved in its writing. The book is available through the Freshwater Habitats Trust for a small donation (see www. freshwaterhabitats.org.uk).

The Wildlife Gardener: Creating a Haven for Birds, Bees and Butterflies, Kate Bradbury (White Owl, 2018) This book is filled with tips and handy advice for making the most of your backyard space. As well as information on garden residents including frogs and newts, Bradbury also details with glee a host of accessory features for your wildlife garden, including bee nest boxes, like those mentioned in this book.

Wildlife Gardening for Everyone and Everything, Kate Bradbury (Bloomsbury Publishing, 2019) This is a useful and easy-to-follow gardening guide with a strong focus on the different types of wildlife you can attract to your garden. Endorsed by The Wildlife Trusts and the RHS, quite frankly, there is no better go-to guide for would-be wildlife gardeners.

IMAGE CREDITS

Key t = top; l = left; tl = top left; tr = top right; c = centre; cr = centre right; b = bottom; bl = bottom left; bc = bottom centre; br = bottom right.

Abbreviated photo sources AL = Alamy; FL = FLPA; GAP = Garden and Plant Image Collection; G = Getty Images; IS = iStock; NPL = Nature Picture Library; RSPB = RSPB Images; SS = Shutterstock.

Front cover SS; **back cover** t Neil Phillips/RSPB; Tom Mason/RSPB; b Anna Williams; **1** Claudia de Yong Garden Designs www.claudiadeyongdesigns.com; **4** l SS, t Nick Upton/AL; **5** bl Anna Stowe Botanica/AL, br SS; **7** SS; **8** Acabashi/CC BY-SA 4.0; **10** Brian North/G; **11** SS; **12** l Mike Read/AL, r Patrick LORNE/G; **13** t De Agostini/Biblioteca Ambrosiana/G, b SS; **14** t Ken Edwards/A, bl Large Birch Creek Pingo/bottomdollar/CC BY-SA 3.0, br Arterra/UIG via G; **15** Christer Fredriksson/G; **16** t Old Images/A , b SS; **17** t Adrian Sherratt/A, b SS; **18** t ART Collection/A, b Ed Crowther Places/A; **19** t Koakoo, VisitBritain/Martin Brent/G; **20** Gerard Puigmal / G; **21** SS, c Manor Photography/A; **23** IS; **24** P A Thompson/G; **26** t GAP Photos/GAP, Kentaroo Tryman/G; **27** tl Rebecca Beusmans/A, tr wda bravo/A, b IS, **28** t John Keeble/G, b IS; **29** t Ron Evans/GAP, b Matthew Doggett/A; **30** t Mark Winwood/G, c SS, b David Squirrel/Flickr/Zippa Pizza; **32** t Heather Edwards/GAP, b Jules Howard; **33** IS; **34** tl SS, tr John Richmond/A, cl Carole Drake/GAP, b Anna Stowe Botanica/A; **35** Ernie Janes/A; **36** tl SS, tr Jonathan Need/A, c Nature Production/NPL, b Mark Winwood/GAP; **37** Willem Kolvoort; **38** IS; **39** t Stephen Barnes/Plants and Gardens/A, bl Colin Milkins/G, br Oxford Scientific/G; **40** Kris Mercer/A; **41** t Jelger Herder/Buiten-beeld/Minden Pictures/G, b SS; **42** Claudia de Yong Garden Designs www.claudiadeyongdesigns.com; **43** Matthew Roberts; **44** Anja Kulovesi-Leino; **46** www.railwaysleepers.com, Nottingham (0115 9890445); **47** www.railwaysleepers.com, Nottingham (0115 9890445), cl SS, cr SS, br IS; **50** SS; **51** SS; **52** Claudia de Yong Garden Designs www.claudiadeyongdesigns.com; **54** SS; **55** SS; **56** tl SS, tr IS, br Radharc Images/A; **57** t VisitBritain/David Sellman/G, c SS, b Jason Smalley/GAP; **58** t Gillian Pullinger/A, b Tim Graham/G, **59** IS, bc Neil Wyatt, br Caterpillar of Cinnabar Moth on ragwort/Leonora (Ellie) Enking/CC BY-SA 2.0; **60** t Francois De Heel/G b Claudia de Yong Garden Designs www.claudiadeyongdesigns.com; **61** l David Chapman/A, tr Courtesy of Lichfield District Council, br Francisco Martinez/A; **62** t Pixel Youth movement/A, b blickwinkel/A; **63** SS; **64** IS; **65** t George McCarthy/NPL, b Cyril Ruoso/ Minden Pictures/G; **66** t Fiona Lea/GAP, b Annie Green-Armytage/GAP; **67** b 10'000 Hours/G; **68** SS; **70** SS, t Charles Krebs, tr Howard Rice/G; **71** Arco Images GmbH/A; **72** l Trevor Chriss/A, c Mark Winwood/G, r IS; **73** GAP Photos/GAP; **75** SS; **76** Vivien Kent/A; **77** t Gwendoline Pain/A, b Kay Roxby/A; **78** t BBA Travel/A, b Ian Alexander/CC BY-SA 4.0; **79** Tim Gainey/A; **80** Daniela White Images/G, t SS, b IS; **82/83** Annie Green-Armytage /GAP; **82** IS, t Jonathan Buckley; **83** t John Swithinbank/GAP, c Mark Winwood/G, b CasarsaGuru/G; **84** IS; **86** IS; **87** SS; **89** Friedrich Strauss/GAP; **90** tl Chris Robbins/A, tr Nature Picture Library/A, bl Donald Hobern, br Frank Hecker/A; **92** SS, tl IS, br Stratiotes aloides/Jörg Hempel/CC BY-SA 3.0; **94** SS, tl Brian Hird (Wildflowers) / A, tr TorriPhoto/G, **96** SS, tr Premaphotos/A, **98** SS, l Benjamin Blondel/CC BY-SA; **99** Field pond/Patrick Roper/CC BY-SA; **100** SS, **102** Alex Hyde/NPL; **103** SS; **104** SS; **105** t blickwinkel/A, b SS; **106** IS; **107** t Adrian Davies/NPL, c Rat-tailed maggot/LoKiLeCh/CC BY-SA, b Yon Marsh Natural History/A; **108** t Michael Durham/NPL, b SS; **109** blickwinkel/A; **110** Minden Pictures/A; **111** tl Picavet, tr Stephen Dalton/NPL, b Chris Lawrence; **112** IS; **113** tl Colin Milkins/G, tr Laguna Design/G, bl Neil Phillips/A, br SS; **114** tl Jan Hamrsky/NPL, bl Andrew Harmer/A, br blickwinkel/A; **115** Willem Kolvoort/NPL; **116** t Imagebroker/FL, b Kim Taylor/NPL; **117** l Buiten-Beeld/A, Nature in Stock/A; **118** t Mike Snelle, b SS; **119** SS; **120** t SS, b Ingo Arndt/Minden Pictures/FL; **121** Andrew Darrington/A; **122** l SS, r Natural Visions/A; **123** tl Natural Visions/A, tr SS, b Joe Blossom/A; **124** t Minden Pictures/A, b Buiten-Beeld/A; **125** SS; **126** SS; **127** t SS, c IS, b Kristian Bell/G; **128** Andrew_Howe/G; **129** SS, bl Amy Lewis; **130** SS, bl IS, br Richard McManus/G; **132** IS; **134** l Alison Stevens, r Yon Marsh/A; **135** Adrian Sherratt/A; **136** Ross Hoddinott/Nature Picture Library/G; **137** Matthew Bigwood/G; **138** l IS, r Dave Bevan/A; **139** l keith burdett/A, r David Chapman/A; **140** Surrey Wildlife Trust/Jon Hawkins; **141** t Auscape/G, b National Geographic Image Collection/A; **142** Quekett Microscopical Club; **143** Ilene MacDonald/A; **144** blickwinkel/A, **145** Jules Howard; **148/149** IS; **152** Matthew Roberts; **153** t Image courtesy of Freshwater Habitats Trust, b Photofusion/G; **154** TNT Magazine Pixate Ltd/A; **155** IS.

The illustrations on pages 48, 49, 51, 88 and 147 are © Bloomsbury Publishing and were commissioned from Wildlife Art Ltd. for *The Wildlife Pond Handbook*.

INDEX

ACKNOWLEDGEMENTS

Ponds have always been a very special place for me and so writing this book has been an immense pleasure. I am enormously grateful to all at The Wildlife Trusts and Julie Bailey at Bloomsbury Publishing for commissioning me to write it. Big thanks also to Alison Stevens for finding such an impressive array of photos and to Austin Taylor for the book's wonderful design.

I am indebted to a number of scientists and other experts whose written words over the years have really helped me gauge and understand further this incredibly rich and diverse habitat. These people include Trevor Beebee, Jeremy Biggs, Ruth Carey, Arnie Cooke, Roger Downey, Tony Gent, Samantha Goodlet, Howard Inns, Tom Langton, Rob Oldham and Froglife's Kathy Wormald. This book owes a huge debt to the author of its predecessor, *The Wildlife Pond Handbook*, Louise Bardsley.

Lastly, a dedication. This one is for the tireless teachers who have inspired students at the pond edge over the years, providing younger generations with positive memories and experiences, and countless springboard moments towards scientific endeavour later in life. All power to these wonderful people.